HERSCHEL GRAMMAR SCHOOL

letter-writing, e-mail & texting

essentials

helen smith

THE LEARNING RESOURCE CENTRE
HERSCHEL GRAMMAR SCHOOL
NORTHAMPTON AVENUE
SLOUGH SL1 3BW

foulsham
LONDON • NEW YORK • TORONTO • SYDNEY

WITHDRAWN

foulsham
The Publishing House, Bennetts Close, Cippenham,
Slough, Berkshire, SL1 5AP, England

ISBN 0-572-02847-4

Copyright © 2003 W. Foulsham & Co. Ltd

Cover photograph © The Image Bank

All rights reserved.

The Copyright Act prohibits (subject to certain very limited exceptions) the making of copies of any copyright work or of a substantial part of such a work, including the making of copies by photocopying or similar process. Written permission to make a copy or copies must therefore normally be obtained from the publisher in advance. It is advisable also to consult the publisher if in any doubt as to the legality of any copying which is to be undertaken.

Printed in Great Britain by Cox & Wyman Ltd, Reading, Berkshire

Contents

Introduction

Letters have proved a great way to communicate for many centuries, from the hand-delivered parchments of old to the multi-coloured missives that drop through your letterbox or into your e-mail inbox today. Writing letters is something that none of us can avoid for ever – and getting it right can make a real difference. A well-written letter can help you win that job, solve that problem, obtain a deserved apology or convey a heartfelt message in words that will be best appreciated by its reader.

These days, of course, the postman brings not only family news and bank statements, but also demands, offers and notifications from councils and utilities; personal requests from companies and charities you've never heard of; news from the residents' association; even pleas for support, signatures or sponsorship. Add to that the messages you receive on your computer or mobile phone, and it seems the whole world is trying to communicate with you – and you with it. Indeed, in this downpour of information, it can be difficult to make yourself heard when the recipient of your letter has so much to wade through each morning.

So in this busily messaging world, how do you ensure that the letter, e-mail or text you send gets noticed, that you put across what you want to say or receive the answers you want? When is it best to use e-mail, and when will only a letter do? Should you write your job application by hand or type it? Should you text your family news or send a letter instead? And how do you make sense of those text messages?

It can often be tricky to get these things right, and that's what this book will help you do. We will look at:

- How and what to write in formal, informal and difficult situations.
- How to get your point across as concisely and effectively as possible.
- When a letter should be hand-written or personally delivered.
- When and how to use e-mail or text messaging instead of a letter.
- How to express thanks, complain, refuse a request or ask someone to stop writing to you.
- Where to get extra help with particular issues.

There are a few subjects that this book does not cover in detail: business-to-business letters, faxes, writing CVs, using or installing e-mail software, text-message abbreviations. For further reading on these subjects, as well as those covered in this book, see pages 183–187.

With our three options of letter, e-mail and text you can communicate effectively in any situation, whether with friends, neighbours and relations, businesses or your local authority. Let's find out how.

Chapter 1

Letter, E-mail or Text?

These days, people write to one another on far more than just paper – with computer and telephone technology we can now communicate in writing almost instantly, anywhere and at any time. There are some situations where only a letter will do; some where e-mail is far quicker and perfectly acceptable; and some where a text is the only practical solution.

The letter

The letter is the best established and most trusted form of written communication. It is the best method for:

- Invitations or announcements – such as weddings, funerals, births.
- Expressing thanks, sympathy, regret or apology.
- Correspondence to family and friends.

It is also the most acceptable in the event of legal proceedings, so is ideal for:

- Formal communication, such as correspondence with a business.
- Disputes or to refute a claim against you.
- Complaints and refusals.
- Answering a confrontational letter you have received.

In fact, there are few situations where a letter is inappropriate. Its only drawback can be the time it takes to write, send and receive. But with today's technology, a simple 'form' letter can be used many times over with little effort, so the writing of many similar letters (such as thank-yous or invitations) is quicker and easier than it used to be. There are usually several options for speeding or tracking the delivery of your letter, depending on the postal service where you live (we will look at this in the next chapter), but if your message simply can't wait until tomorrow to be delivered you'll need another option.

The e-mail

E-mail has been available for 30 years but only with the spread of the internet has it become popular. Many people now have daily access to e-mail either at work or at home, and those that don't can use the facilities at internet cafés and libraries. It is a quick, easy-to-use method of communication, particularly when writing to more than one person. It's best used for:

- Fast, direct communication anywhere in the world.
- Chatty communications with family and friends.
- Disseminating information, documents, pictures and video.
- Keeping in touch with a group of people.
- Simple requests to named departments of businesses/councils.
- Confirmation of information or requests.

However, in most cases e-mail is considered both insecure (that is, it can be read by someone other than the addressee) and informal, so it should not be used for:

- Sensitive situations, e.g., conveying sympathy in bereavement.
- Apologies for anything other than a trivial offence.
- Situations with potential legal consequences (e.g., a dispute over payment).
- Conveying confidential, personal or sensitive information.
- Telling your friends your holiday plans.

As you will need the precise e-mail address for the person you wish to reach, e-mail isn't ideal for initial complaints or enquiries to, for example, a department at the council, unless you are able to obtain the correct address from another source, such as the council's web site. Also, it can be difficult to prove whether a particular e-mail message has been received by its intended reader, so in the case of a demanding or contentious message it is all too easy for the addressee to blame the computer and say it never arrived. So if it is essential that your message gets through, use registered mail instead.

The text message

Text messaging (sometimes called texting or SMS) is a great way to get a short message from your mobile phone to someone else's quickly and discreetly, without interrupting them with a telephone call. (Some new telephone products, such as Amstrad's e-m@iler™, allow you to send texts to a mobile but not to receive them back, although things are bound to change as technology develops.)

While texting is certainly not a replacement for the letter, being particularly informal and often abbreviated (you can use only 160 characters including spaces), it can be used effectively in many situations. For example:

- Telling someone you'll be late.
- Arranging a meeting or night out.
- Sending a few lines of information.
- 'Chatting' when you can't make yourself heard, e.g., in the cinema.
- Confirming in writing what you've discussed on the phone.

Like e-mail, texting is not always entirely reliable or secure, so should only be used for personal messages and never for anything confidential or sensitive in nature. You should never use text messaging for:

- Complaints.
- Expressing sympathy or condolence.
- Invitations or replies – the messages are too easily lost.
- Anything with legal consequences.

> IMPORTANT!
>
> Never use mobile telephones in hospitals, on aircraft, in fuel station forecourts or other restricted areas. This is because the signal they emit can interfere with electronic equipment, including medical monitoring devices. This also applies to laptop or palm computers using infra-red or wireless internet connections. Never use any kind of communication device when driving or operating equipment.

Some examples

Here are a few examples of which method to use in particular situations. Use this to help you decide which method is best for what you want to write. Of course, sometimes you'll be in a hurry and simply need to phone! In some circumstances, the best choice will depend on the person you are communicating with. On the whole, older people tend to prefer to receive a thank you letter rather than an e-mail, for example, while others will be quite happy with a text. Use your discretion.

Correspondence or the phone book are the best places to find addresses; web sites are a good source of business e-mail addresses. Remember that e-mail is not advisable if you are including information that you want kept secure (when you are going to be away on holiday, for example), although you can use it if the message is written cryptically.

	Letter	E-mail	Text
At home, I want to ...			
Query a charge on my electricity bill	✔	✘	✘
Ask my landlord to repair a leaking gutter	✔	✔	✘
Explain to my landlord why the rent is late	✔	✘	✘
Ask my flat-mate to pay the rent I'm owed	✔	✘	✘
Ask the council for advice on planning permission	✔	✔	✘
Ask a company to provide a quotation for building work	✔	✔	✘
In my social life, I want to ...			
Change an appointment to meet a friend	✔	✔	✔
Complain about lousy service at a restaurant	✔	✘	✘
Book a hotel room or enquire about facilities	✔	✔	✘
Reply to a letter	✔	✘	✘
Reply to an e-mail	✔	✔	✘
Ask to be removed from a mailing list (paper or e-mail)	✔	✔	✘
Ask a friend to return an item I loaned them	✔	✔	✔
Invite friends for dinner	✔	✔	✔
Invite guests to my wedding (or similar formal event)	✔	✘	✘
Refuse an invitation	✔	✔	✘
Send my address/phone number to somebody	✔	✔	✔

	Letter	E-mail	Text
Tell my friends about my holiday plans/flight details	✔	✘	✘
Express sympathy over death or illness	✔	✘	✘
Broadcast good news (e.g., birth of a baby)	✔	✔	✔
Ask a friend if I can stay for a few days	✔	✔	✘
Send news to family and friends	✔	✔	✔
Thank someone for a gift/their hospitality	✔	✔	✘
Pass on a message or information	✔	✔	✔
Communicate discreetly with someone in a quiet place	✘	✘	✔
Get a message through in a noisy place	✘	✘	✔
About money, I want to ...			
Ask for a bank loan or statement	✔	✔	✘
Ask my financial adviser to check up on a policy	✔	✔	✔
Accept a quotation and ask for work to start	✔	✘	✘
At work, I want to ...			
Change an official or business appointment (unless told e-mail is acceptable)	✔	✘	✘
Provide a reference for a friend seeking a job (unless told e-mail is acceptable)	✔	✘	✘
Ask a friend to provide a reference for me	✔	✔	✔
Ask someone to explain something that's confusing me	✔	✔	✘
Apply for a job (unless told e-mail is acceptable)	✔	✘	✘
Accept/reject a job offer (unless told e-mail is acceptable)	✔	✘	✘
Resign from my job or a position of leadership	✔	✘	✘
Complain to my employer about facilities/working practice	✔	✘	✘
Ask for time off (e-mail may not be appropriate)	✔	✔	✘
Tell my colleagues we've won that important contract	✔	✔	✔
Send samples of work to prospective employers/customers	✔	✔	✘
Ask for a loan to be repaid or for goods I supplied to be paid for	✔	✘	✘

Getting it right

Whichever method you choose, there are some things that always make a difference to how your message is received.

Layout: If your message is not clearly laid out, with all the usual bits and bobs in all the right places, in neat writing or a clear typeface and with sufficient space between lines and paragraphs, your reader will find it very frustrating and may just cast it aside.
Structure: Getting your thoughts in the right order and getting your point across is all important, particularly when writing to businesses or when you want something to be done as a result.
Politeness: No matter how much you may be seething, a little politeness goes a long way to getting what you want. If you need to criticise, do so dispassionately, not with venom. If you want something done, remember to say please and thank you – and always address people in the way they prefer.
Spelling: Okay, so few of us spell perfectly, but it does make a difference, especially when you are trying to impress, and particularly for job applications. These days it's easy to check your spelling; use a dictionary or, if you are using a computer, the spell-check facility – although be aware that the latter is not foolproof. Always read your message through before sending, and remember: while abbreviated words can be used in text messages it is essential to do so correctly or your message may be misunderstood.

We shall look at these items in detail in Chapter 2. Meanwhile, let's look at some other things you need to think about before getting started.

The purpose of your communication

Before you start, think about why you are writing. Is it to:

- Ask for or provide information?
- Ask for something to be done or considered?
- Demand action (usually after an earlier request has failed)?
- Acknowledge or celebrate an event?
- Give bad news?
- Decline a request or invitation?

Most letters have one or more of these purposes, and the way in which you plan your letter will depend on this. For example, when applying for a new job you will need both to provide information to the recipient showing that you are appropriate and ask them to consider you. Let's look at an example.

Mr Brown has discovered a possible error in his bank statement. He is writing to ask the bank to check and correct the problem. He needs to:

- Provide information: his bank account details and details of the possible error.
- Request action: ask the bank to check and correct its records.
- Acknowledge: thank the bank clerk for looking into the matter.

Notice how the paragraphs are short and to the point. Mr Brown states his purpose – to get the error corrected – then details the issue, with reference to the evidence attached. He does not accuse or ask for an apology but gives the clerk all the information she requires to investigate and resolve the situation. He provides his contact number in case of further query, and thanks the clerk for her help.

26 Nutley Lane
Hanford
West Sussex
BN26 9HW

28 June 2004

Premier Bank
High Street
Hanford
West Sussex
BN26 8TN

For the attention of Joan Williams

Dear Ms Williams

Re: Current Account 19287462, Mr John W. Brown

I have received my May statement and believe it contains an error that has resulted in unnecessary charges being levied to my account. Would you please check the details and make the necessary corrections?

On 2 May I made a postal deposit of two cheques, one for £480.26 and another for £25.63 – a total of £505.89. A copy of the deposit counterfoil is attached.

On my May statement (copy attached) this deposit appears as £50.59 (6 May). As a result, my account has become overdrawn and attracted interest charges.

Please would you check your records and call me with your assessment of the situation. You can contact me during office hours on 02836 451362.

Thank you for your assistance in this matter. I look forward to hearing from you within the next few days.

Yours sincerely

John Brown

John Brown

Enc. Deposit counterfoil no. 29372; May statement from the above account

Consider your reader

Think about the person who will receive your letter. This will affect the way you write and the language you use. For example:

- An elderly person would expect more formal language, especially from someone they don't know very well.
- A young child will eagerly read letters if written clearly in large letters using simple words.
- Certain words and expressions have legal or financial connotations, so choose your words carefully.
- A foreigner may have limited command of your language, so beware of using clichés and idiom that they may not understand.
- Someone who is upset or emotionally charged may take offence easily, so take extra care with the phrasing of apologies, condolences and sympathy letters to avoid unintentional slip-ups.
- If your reader is expecting a different response than you are able to give, let them down gently using calm, persuasive language.
- Unless you know the person well, don't use humour at their – or anyone else's – expense.
- If you are writing on behalf of an organisation, keep your own views and thoughts out of the letter to avoid confusing your reader.
- If your letter contains personal information, mark it and the envelope as such to save any embarrassment.

Getting your points across

There are several things you can do to make your letter easier to read and understand:

- Use short paragraphs (no more than five lines) and sentences (no more than 20 words).
- Use basic language, not unusual words (unless necessary).
- Avoid clichés and outdated phrases (such as 'at this point in time').
- Use plenty of space between lines and paragraphs.
- Use a simple typeface or write neatly.
- Check your spelling and punctuation. Don't rely on the computer's spell-checker – it won't pick up 'an' in place of 'and' or 'through' in place of 'thorough', for example.
- Put information in a logical order.
- Speak to your reader – try to use 'you' rather than 'I' or 'we'.
- Be polite, no matter how angry you may be.
- Get to the point – don't waffle!
- Give contact details in case the reader needs further information.
- Repeat any requests for action at the end of the letter, giving a deadline where possible.

Things to remember

- **Letters:** These are best for long messages, for formal or business communication and for invitations and announcements.
- **E-mails:** These are best for chatty communications, for quick requests and for messages to more than one person.
- **Texts:** These are best for 'conversations', for last-minute changes of plan and for discreet communications anywhere at any time.
- Think about the reasons for your message and what you hope to achieve with it before you start to write, and consider what your reader will expect when they receive it.
- Always be polite, consider structure and layout, and check your spelling.

Chapter 2

How to Write a Letter

A letter is more than just a collection of words and sentences. It needs a beginning, a middle and an end. It needs to show who it is going to and who it is coming from, as well as when and why, and it needs to be appropriately worded and styled for its purpose.

Depending on where your letter is going and to whom, there are different styles of presenting the information. Sometimes an informal approach is fine, while other letters require a businesslike tone and professional layout if you are to be taken seriously. Let's look at the available options.

Formal and informal letters

Letters can be formal or informal in style and tone. For example:

Formal letters include:	Informal letters include:
Letters to businesses and authorities	Letters to family and friends
Letters of complaint	Thank you notes
Letters threatening action	Letters of apology or condolence
Invitations to formal events (weddings, funerals, gala dinners, etc.)	Invitations to informal events (parties, sports/charity events, etc.)

Some of the items that must appear in a formal letter can be omitted from an informal one. For example, a formal letter must always show the name and address of the person to whom you are writing, but this is unnecessary on an informal note.

Informal letters can be more chatty, reflecting the way you speak, rather than requiring the rigid adherence to the rules of grammar and punctuation that is more common in formal correspondence. For example, it is acceptable to use 'don't' and 'can't' in informal letters, while the formal would require 'do not' and 'cannot'.

Should I write or type?

Many of us now have the opportunity and ability to use a typewriter or computer to produce letters. But when is it acceptable to do so?

- If possible, always type formal letters, particularly those of complaint or concerning money, dates or other figures that could be misinterpreted if hand-written. Letters to businesses/local authorities may need to be photocopied, and a typed letter produces a clearer image. Always use 'standard' typefaces such as Times or Arial for these letters.
- If you're writing the same letter to several people at once, typing it will save a lot of time. Many word-processing programs have templates for items such as invitations and newsletters, including appropriate pictures or borders and a range of typefaces and colours to make them more eye-catching. You can even include pictures of your family if you have a scanner.
- Always write personal letters, including letters of sympathy or thanks, by hand. This shows you have taken the time to consider

your words rather than just banging out a copy on your word-processor. Take your time and make them neat. Whenever possible, deliver by hand; it shows you have taken time and thought.

- With many jobs now requiring keyboard skills it is acceptable to type a job application, as this shows off your IT skills. However, if the job requires you to write large amounts of information by hand (for example, a shorthand secretary or a court clerk), a hand-written application is preferred. This is usually indicated in the job advertisement, but if in doubt, ask the personnel department for advice. Always type your curriculum vitae, taking care over the layout, and present it on good-quality white paper.
- Whether writing or typing, use only one side of the paper unless writing a chatty letter to a relative or friend. If you use more than two sheets of paper, number them clearly from page 2 onwards.

What paper should I use?

These days even computer paper comes in a dazzling array of colours and styles. Which is best for letters?

- For formal letters and job applications, always use plain white paper, A4 (in Europe) or Letter size (in the US and Asia), and pay attention to layout (see next section). If possible, use paper of 90-110gsm (this will be printed on the packet) and do not choose a 'weave' design as this will impair the print quality. The ideal paper for use in computer printers is labelled 'Bond' or 'Script'.
- For short letters, it is sometimes better to use A5 paper (which is the same size as a sheet of A4 folded in half). This looks better than a few lines of print 'drowning' in a sea of white space.

- For invitations and personal letters, A5 is more usual (sometimes smaller for invitations). Thick paper can be used to create cards if your printer will take it. If using coloured paper, do a test print first to check the colours appear on the coloured background correctly. Remember that some combinations can be difficult to read (red ink in particular).
- For letters of sympathy always use plain, good-quality paper with no decoration and a pen that writes smoothly in black or blue ink.
- Blank cards or stylish notelets are ideal for thank you notes and short personal letters, but you shouldn't cover every surface with writing. If you have a lot to write use good quality writing paper in a plain colour, preferably white, cream or air mail blue.
- It is no longer necessary to write international letters on air mail paper, but it does reduce the cost of the postage.

What about envelopes?

There used to be a certain etiquette attached to the use of white or manila/brown envelopes, but this has largely disappeared. Generally:

- Letters to businesses or job application letters may be better received in quality white envelopes.
- Letters of condolence or expressing bad news should be sent in plain white envelopes.
- Decorated envelopes should be used only for friendly, personal correspondence.

Choose the size of envelope carefully. Don't use an envelope that's too big and that allows the contents to move about.

- Most formal letters can be contained in a DL size envelope, which is one-third the size of a sheet of A4 paper.
- If your letter is bulky – say, if you are including a document such as your CV – you may prefer to use a larger envelope, such as A4.
- Use a padded envelope if the contents are delicate or unusually shaped and need protection. These are now available in manila, white and other colours and in a wide range of sizes.

We will look at how to address envelopes later in this chapter.

Structuring a letter

The following structure should be used for all formal letters. Informal letters can omit those items shown in italic, but the same formats should be used for addresses, etc. Note the use of white space in the example overleaf. Make sure your own letters use white space in this way to avoid a cluttered appearance.

1 **Your address:** Always show your full address with the post town and postcode in capital letters. If the county name is the same as the town, omit it (e.g., Gloucestershire should not follow Gloucester). If your letter is going to another country, add the name of your country at the end of your address after the postcode.
2 *Reference code:* If replying to a letter, quote the reference number or code on that letter. Otherwise omit this line.
3 **Date:** Always write dates with the month written out in full and include the year. Do not write 'Date as postmark' as this is useless if the envelope has been discarded. In the UK dates are normally written in the form 23 April 1967; in America and Canada this would be April 23 1967.

RESOURCE CENTRE
HERSCHEL GRAMMAR SCHOOL
NORTHAMPTON AVENUE
SLOUGH SL1 3BW

42 Botley Close
① Colebourn
YORK
YK18 7QS

Your Ref 5/12A ②

17 May 2003 ③

Without Prejudice ④

Special Delivery ⑤

Brown's Drives
28A Long Lane ⑥
YORK
YK12 1AN

FOR THE ATTENTION OF MR JOHN BROWN ⑦

Dear Mr Brown, ⑧

Completion of work at 42 Botley Close ⑨ ⑩

Thank you for your letter dated 5 May 2003.

I realise that you have had staffing difficulties and that this has affected your work schedule. However, this is not sufficient reason for you not to complete the work as agreed, especially as we have already paid you. I therefore advise you that if the matter is not resolved by Friday next we shall be instructing our solicitor to pursue a claim against you for breach of contract. We have not taken this decision lightly but as you are aware it has now been four months since you started and we are still unable to use our driveway.

I look forward to seeing you as soon as possible to finish the job.

Yours sincerely, ⑪

John Smith ⑫

cc: Mr F. Hanson (Hanson, Little and Wright Solicitors) ⑬

4 *Classification:* This is optional but should appear if your letter is sent to a business address but contains personal or confidential information. 'Personal' indicates the letter must only be opened by the person to whom it is addressed; 'Confidential' indicates it can be opened and read by someone else (e.g., a secretary), but that the information it contains must be kept secret. 'Without Prejudice' is often used on solicitors' letters and is a legal term indicating that the text of the letter does not diminish the rights of or admit liability by either party to the letter. If you use a classification, show it on the envelope as well.

5 *Despatch method:* Usually unnecessary but if timing is important, show the method used to speed the letter along (e.g., 'Special Delivery', 'By Courier', 'Registered Mail').

6 *Recipient's name and address:* Wherever possible address your letter to a person, rather than a department. If you can't obtain a name, give a job title (for example, 'Personnel Manager'). Make sure you use the correct, full address, including any room or box number, and put the post town and postcode in capital letters without punctuation. UK postcodes can be found on the Royal Mail's web site www.royalmail.com/paf or by calling 08457 111222.

7 *For the attention of ...:* In letters to businesses or authorities, you can place the recipient's name on this line rather than showing it as part of the address. It should be underlined or shown in block capital letters with no full stop, for example: 'FAO John Jones, Personnel Manager' or 'ATTN: JANICE WRIGHT'.

A note on names

If you are replying to a letter, address the person as they have shown their name. For example, if the letter to you is sent from Mr P J Jones, write this at the top of the address and start the letter 'Dear Mr Jones'; if the letter you received was signed 'Phillipa Smith MD' write this at the top of the address and open the letter with 'Dear Dr Smith'. There are rules for addressing people with specific titles and roles, such as doctors, the judiciary, religious leaders, royalty, knights and government officials: for details see 'Addressing titled persons' on pages 105–111.

8 **Opening greeting (salutation):** It is important to get this right, particularly for persons of title. General guidelines are shown below, but for titled or professional persons there are specific rules, as discussed above, so make sure you use the correct form.

Letter to:	Start with:
Someone whose name you don't know, or whom you are addressing in their professional capacity	'Dear Sir', 'Dear Madam' or 'Dear Sir or Madam' (if unsure)
An entire company or department	'Dear Sirs' or 'Dear Mesdames' if all staff are female, otherwise 'Dear Sirs'
A named individual whom you do not know personally	'Dear Mr Jones' (or Mrs, Miss or Ms), or 'Dear Janet Smith'
A colleague, relative or friend	'Dear Arthur' or 'Dearest Angela'

If in doubt, the best guide is to think how you would address that person if you met them in the street, and use that form of their name.

9 *Subject heading:* Use only for business letters. This gives the recipient some idea of what the letter is about. If replying to a previous letter, use this line to refer to that correspondence (e.g.,'YOUR LETTER OF 12 JUNE RE: 24 BROOK LANE' or 'Delivery of goods for order 19234/27'). Either underline or type in block capitals with no final full stop.

10 **Main text:** We will look at how to structure the main text of your letter in the next section. It is particularly important in business letters to get the opening and closing lines right and to state your purpose clearly and concisely. For personal letters this is less important but it is a good habit to adopt.

11 **Complimentary close:** For formal letters, if you have addressed the letter to a specific person who you have met or spoken to at length, use 'Yours sincerely'; otherwise 'Yours faithfully' is more appropriate. In some countries 'Yours truly' is preferred for all correspondence. For titled or professional persons there are specific rules, as for opening and address lines (see pages 105–111). In personal letters most of us use 'With kind regards', 'Best wishes', or something less formal, but you should ensure they match the context of the letter. 'Best wishes' would be inappropriate in a condolence letter, for example.

12 **Signature and sender's name:** If signing on behalf of a business, you can add the company name beneath your signature and name. Otherwise show your name in the form you prefer to be addressed, as this will be used for any return correspondence. Ladies – remember that if you show only your initials or have an unusual

name it will often be assumed you are male; so if you prefer to be addressed 'Miss', 'Ms' or 'Mrs' make sure you show this. For informal letters a signature will suffice, as long as it is recognisable!

13 *Enclosures and copies:* If you are enclosing documents note them here with the prefix 'Enc:'. It helps you to remember to send the items and the recipient is able to check they are present. If your letter is copied to several people, note their names here so the recipient does not need to send copies to them. Use the abbreviation 'cc' or 'Copies to' to indicate this.

How to lay out a letter

As computers and typewriters have become more widely used, the layout of letters has evolved. No longer the need for elaborate indents and tab stops: most letters now follow the fully blocked format where all items are set to the left-hand side of the page – except the sending address, which is set in a column on the right unless headed paper is used – and paragraphs are separated by a double line space. In the following example, note how there are no commas after lines of the address, or full stops within abbreviations. Dates are shown '30 June' rather than '30th June'. The use of double spaces between paragraphs helps to add white space, which breaks up the text and makes for easier reading.

68 Constable Road
Gainsborough
IPSWICH
Suffolk
IP18 7QS

17 May 2004

Jane's Arts and Crafts
High Street
HANDLEY
Devon
EX12 1XY

FOR THE ATTENTION OF MRS JANE CHEADLE

Dear Jane

Thank you so much for processing my recent order so quickly. I have received all the items and am very happy with them.

However, the receipt I have received does not match the amount taken from my credit card. I expect there was an additional charge for postage; could you please confirm this and send a revised receipt as the amount (£24.68) was more than I had expected. Please enclose a copy of your latest catalogue as well, as some friends have expressed an interest in purchasing from you.

I look forward to hearing from you soon.

Yours sincerely

A.J. Ramford

Angela Ramford

There are two other layouts in common use. First the 'semi-blocked' style:

68 Constable Road,
Gainsborough,
IPSWICH,
Suffolk,
IP18 7QS

17 May 2004

Jane's Arts and Crafts,
High Street,
HANDLEY,
Devon,
EX12 1XY

For The Attention Of Mrs Jane Cheadle

Dear Jane,

Thank you so much for processing my recent order so quickly. I have received all the items and am very happy with them.

However, the receipt I have received does not match the amount taken from my credit card. I expect there was an additional charge for postage; could you please confirm this and send a revised receipt as the amount (£24.68) was more than I had expected. Please enclose a copy of your latest catalogue as well, as some friends have expressed an interest in purchasing from you.

I look forward to hearing from you soon.

Yours sincerely

A.J. Ramford

Angela Ramford

As you can see this uses more punctuation; paragraphs in the main body of the letter are indented and the signature appears in the centre.

The fully indented layout has stepped addresses, reduced space between paragraphs and the signature has moved to the far right.

68 Constable Road,
Gainsborough,
IPSWICH,
Suffolk,
IP18 7QS

17th May 2004

Jane's Arts and Crafts,
High Street,
HANDLEY,
Devon,
EX12 1XY

For the attention of Mrs Jane Cheadle

Dear Jane

Thank you so much for processing my recent order so quickly. I have received all the items and am very happy with them.

However, the receipt I have received does not match the amount taken from my credit card. I expect there was an additional charge for postage; could you please confirm this and send a revised receipt as the amount (£24.68) was more than I had expected. Please enclose a copy of your latest catalogue as well, as some friends have expressed an interest in purchasing from you.

I look forward to hearing from you soon.

Yours sincerely

A.J. Ramford

Angela Ramford

How to address an envelope

An envelope gives the first impression of your letter, so:

- Always type or write neatly. Do not use block capitals other than for the post town and postcode.
- Always use a courtesy title – Mr, Mrs, Ms, Miss (for unmarried woman or girl under 16) or Master (for boy under 16); or other appropriate title such as Rev, Dr, Sir, etc. For details of how to address letters and envelopes for persons of title see pages 105–111.
- Write the name and address lengthways along the envelope with the opening to the top or left. Start about halfway down and one-third of the way across, leaving sufficient room for postage stamps. You may prefer to use a sticky label instead of writing directly on the envelope, particularly if you are using a large or padded envelope.
- Clearly mark any classification ('Personal', 'Confidential') or despatch instruction ('Airmail', 'Special Delivery') in the top left corner.
- If your letter is to be delivered by hand, a name and first line of address, or company name, is normally sufficient. Add 'By Hand' towards the bottom of the envelope, between the centre and right-hand edge of the page.
- For international correspondence, parcels or larger envelopes, include your own name and address on the flap of the envelope. This can appear in a block or as a single line with the elements separated by commas. You may need to complete and attach customs forms for international correspondence.

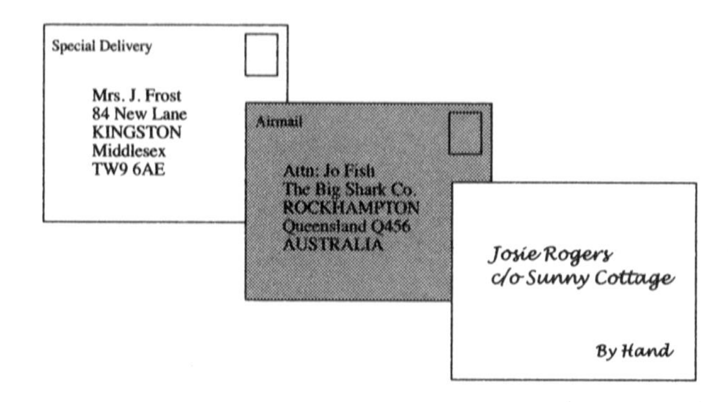

Getting the words right

Now you know how to structure and lay out your letter and envelope. You can use the samples at the back of the book for ideas about how to phrase your questions or make an effective complaint, but you might still be unsure how to write well. Here we look at the main pitfalls and how to avoid them.

Spelling

No matter how good we are at spelling, some words will always be bugbears, and those not in daily contact with a keyboard have the additional problem of mistyping. So it is always essential to check your spelling, particularly on important letters on legal, financial or confidential matters.

Most word processors come with spell-checkers, but how reliable are they? Well, this sentence would have passed the test:

Through she wad pill, high envoyed he pasty.

What on earth was that? Well, it should have been:

Though he was ill, Hugh enjoyed the party.

Because all the mistyped words were acceptable words, the spell-checker did not pick them up as wrong. So be careful! While this is an extreme example, the occasional 'on' for 'no', 'on' for 'in' or 'through' for 'though' (or 'thorough') can easily slip through and ruin your carefully crafted letter.

Use a dictionary for any words you are unsure of, and make sure you read through your letter before you send it.

Grammar

Don't get too hung up on perfect grammar. It's not an exact science anyhow, and if you worry overmuch about dangling participles and split infinitives your letter will sound more stilted than ever. Write how you feel comfortable, but bear in mind the following:

- The words me, myself and I cause many problems. Which is correct: 'from Sheila and I' or 'from Sheila and me'? In fact it is the latter. The easiest way to work it out is to remove the 'other person' (that is, Sheila, in the above example): 'from I' would make no sense, so the correct version must be 'from Sheila and me'. How about 'John is cleverer than me'? In fact this is incorrect: in its full form the sentence should read 'John is cleverer than I am' ('me am' would be nonsense), so 'John is cleverer than I' is the correct form.

- Verbs are singular or plural depending on their subject. For example: 'John and Kathy were in the canteen' but 'John, as well as Kathy, was in the canteen'.
- Adverbs (which generally end in 'ly') describe verbs; adjectives describe nouns. So if you're describing how someone did something, use an adverb ('she read the letters very quickly' or 'she quickly read the letters'); if describing the thing itself, use an adjective ('she read the many letters' or 'she read the congratulatory letters').
- Prepositions can be tricky. Don't use 'due to' when you mean 'because of' or 'following' when you mean 'after' or 'as a result of'.
- Fewer and less: the former can be used for items that can be counted ('Fewer people were on the train this morning'); less for quantities ('There was less fog than yesterday').

Punctuation

Since so many of us seem to lack the basics of punctuation, you may well ask, 'Why bother?' Well, here's an example of why we do need to bother if we want to be sure our reader understands our intentions:

> Consider the sentence 'A woman without her man is lost.'
>
> Which does the writer mean?
>
> A woman: without her, man is lost.
>
> or
>
> A woman, without her man, is lost.

There are many books out there dedicated to teaching you the art of perfect punctuation, so if you need further help, see the list at the end of the book. But here are a few tips for avoiding common mistakes:

- Apostrophes indicate two things only: possession (that is Martin's ruler – it belongs to Martin) and elision (that's my ruler – that is). Do not use them in plurals or for any other word ending in 's'.
- All sentences end with a full stop. If a sentence introducing a list is incomplete, use a colon.
- Also use colons before a quotation or to separate two different subjects within a sentence.
- Use semi-colons to separate clauses in a sentence, where a comma would not provide a long enough pause.
- Use parentheses (brackets) or dashes to enclose a thought or phrase that is aside from the main thrust of the sentence. If using dashes, leave a space either side of each dash. For example: 'Thank you for the wonderful party – you went to a lot of trouble – and the presents you all bought for us.'
- Use question marks only for direct questions ('Are you going to Julia's party?') not for indirect reporting of a question ('She asked whether I was going to Julia's party.').
- Use single exclamation marks only to indicate excitement or surprise. You rarely encounter these marks in a business-related document, but they are often used to excess in family letters.
- When reporting someone's exact words, use quotation marks. Either single (He said 'Can you show me the way?') or double (He said "Can you show me the way?") can be used. Question marks, commas and full stops should appear within the quotation marks where they are inferred by the text (as above).

- Use capitals only at the start of a sentence or for proper nouns that name specific people, businesses, buildings or objects (for example, 'Nelson's Column' but 'the column in Trafalgar Square'; 'the Minister of State for Social Security' but 'various ministers of state'). Days of the week and month names should also have an initial capital. Don't use block capitals other than in specific parts of your letter (see the sections on structure and layout for details).

Things to remember

- Use formal style unless writing to family and friends.
- Write neatly or type – the latter is best for business letters – on one side of the paper.
- Use appropriate paper and envelope for each letter.
- Use the elements of structure that will help your reader understand your letter and respond to it.
- Decide on a layout and stick to it.
- Use plenty of white space, but don't let your text drown on a page that's too large.
- Make sure your letter arrives by addressing it neatly and correctly.
- Use correct grammar, spelling and punctuation and check it all before sending.

Chapter 3

Personally Speaking – Writing to Family and Friends

Writing to family, friends and others we know well is usually a pleasure. It is informal; we can be ourselves and waffle on at our leisure. Then there are invitations and announcements, thank-yous and well-wishing: none too onerous a task, but often etiquette demands a certain style. In this chapter we look at them all, with examples and tips to help you get them right.

Invitations

Everyone loves a party, so invitations are usually fun to write. With so much pre-printed party stationery around, there's little need for you to design and write your own unless you want to. The same can be said for weddings, birth announcements and birthdays.

But what about the funeral, the unusual event, surprise party, or bring-a-dish party? How do you address these? Here are a few tips and examples; you can find other examples at the end of the book.

- Keep it short but give everything they need to know – time, place, names, date, dress code.

- For fun events, use fun invitations: colour, graphics and fancy fonts. For more sober or formal occasions, stick to black with a plain border and use stylish typefaces.
- Use appropriate language for the event; see the examples at the end of the book for ideas.
- Don't be ambiguous with the dress code: if people can wear jeans, say so; if only ball gowns are acceptable, state 'Formal wear ONLY'. This stops all those anxieties about what is appropriate.
- Include an RSVP slip or ask for responses by phone. Always give a deadline.
- If it's a surprise party, say so! Otherwise it may not be a surprise for long.
- If you need people to bring something (a bottle, for example), say so. A separate note is fine if only a few need to do so – but make sure you fold it inside the invitation so they don't miss it.
- If it's a fundraiser, make this clear – especially if there is an entrance fee or if guests will be asked for donations. This saves the embarrassment of guests arriving cashless.
- If you don't want gifts, flowers or whatever, say so. If you would rather nominate a charity to receive any donations, give the name, but there is no need to elaborate.
- If it's a child's party, get the child either to colour the invitations or to write the names. It helps them to feel involved.

There are several examples of invitations on pages 112–117, covering weddings, birthdays, surprise events and formal events.

Announcements

These often take a similar form to invitations, and are often written on pre-printed stationery. Certain announcements, particularly births, engagements and deaths, may also appear in the press – the wording would be similar but may also include thanks or additional details.

Here are some ideas for a more personal approach:

- For birth announcements, include a scanned image of the baby with all the usual statistics. If it's not your first, include the names of the others (for example, a new sister for Amy and Ben). Include thanks to the delivery team and details of the ward on which the new mum and baby can be visited (if you want visitors!) and visiting hours.
- If you don't want gifts and cards, say so – otherwise people may feel obliged to send something. Suggest donations to charity instead.
- For engagements, a photo of the happy couple often goes down well. An informal or semi-formal tone is ideal. You don't have to give details of when or if a wedding is planned. The announcement should be made by those who will be hosting the wedding – for example, the prospective bride's parents, or the couple themselves if older. Similar wording should be used as on wedding invitations – see pages 113–114 and the wedding references listed on page 185 for details. In some places, particularly the US, it is common to include details of the engaged couple's occupations in both announcements to the press and those mailed to family and friends.
- For a change of address/new home announcement, make sure to include all elements of the name and address, plus phone numbers. If a detail has not changed (for example, your e-mail address), say so, but give it anyway.

- If you combine the announcement with an invitation (say, to an engagement party), make this clear! It is often better to send the two separately a few days apart.
- For less happy announcements, such as divorce or death, use a more formal tone. You don't need to give the cause or place of death; e.g., if it was unexpected then 'suddenly' will do fine. Give details of the funeral if it is already arranged and if you want all those receiving notification to attend. If the funeral is to be for close family or invited guests only, say so.
- Funeral announcements should give full name, date of death, date of funeral, time and place, and any instructions on dress, flowers or donations (e.g., family flowers only; formal wear; donations to BHF).
- Keep all announcements short – stick to the facts. You can always include the announcement card in with a longer letter.

The round robin

Despised by some, loved by others, the 'round robin' or 'circular' letter is flourishing, thanks to word processing and the ease of printing many copies of one letter. Often sent with a Christmas card, it aims to inform and entertain, to tell those we correspond with infrequently what's been going on with our family recently. But it's all too easy to get it wrong. Here are some tips:

- Be aware of family feuds or personal dislikes. Don't fan the flames by showing favouritism!
- Keep it short, five paragraphs of four lines is quite sufficient.
- Remember to include at least one detail about each person in the family – including yourself! Try to keep a balance.

- Keep it positive – don't dwell on illness, injury or death, but acknowledge them when appropriate.
- Don't over-apologise for its mass-mailed format – a quick one-line explanation at the most.
- Don't photocopy your letter – it looks cheap. Make the most of its computer-generated layout with colour, images and fun fonts.
- Keep it interesting – don't go into great detail on something few people will want to know.
- Don't send such letters to those who don't appreciate them; write if you've time, otherwise a short note explaining how busy you are will suffice.

You can use almost any stationery for this. Just remember to set your word processor and printer to the appropriate size, as most decorated formats are smaller than A4.

Here are some extra ways to make it personal and different:

- Use mail merge to address each copy personally, rather than saying 'Dear all' – or write the names in by hand.
- Use seasonal or apt graphics to jazz it up.
- Sign each copy personally and ask the rest of your family to do so.
- Write – not type – a short message at the bottom of each note that is relevant to the reader (for example: 'Congratulations to Eddie for his great results – looks like another Smith destined for the Bench!').
- Attach photos of significant events since you last wrote.
- Get the kids to draw pictures on the paper before you put it in the printer.

You can find an example on page 118.

Thank you!

Everyone likes to be thanked, especially when they've made a special effort. Not enough of us take the time to thank people other than for birthday, wedding or new baby gifts – and some don't even bother with that, sadly. But we should – it's a great way to get someone on your side and to make them feel good! If we all took the time to thank those that gave good service in hotels, restaurants and businesses, maybe more people would make the effort.

To really make a good impression:

- Write by hand, unless you've many people to thank. If typed, write your name and a brief personal message at the end.
- Wherever possible, make it personal, naming the gift received and commenting on it, rather than just thanking them for 'the present'.
- Use quality paper, notelets or a card unless thanking someone in their professional capacity.
- Don't go overboard – over-effusive thanks can sound false.
- Repeat your thanks at the end of the letter.
- For wedding/birthday thank-yous, thank them for attending (or for their kind thoughts if they didn't attend) as well as for the gift. If the person was unable to attend the event, give a brief description.
- For new baby thank-yous, include a picture of the baby and one of the new family. You are expected to give all the usual details and sound enthusiastic – even if you're exhausted!
- To thank someone for attending a funeral or memorial service, a printed card is ideal. Acknowledge any donation or flowers and thank them for their kind thoughts. Be brief.

- When thanking someone for their help or for good service, consider sending a recommendation and summary of the situation to their boss – it always helps to get their effort recognised and encourages them to do the same for others.

You can find examples of thank you letters on pages 119–123.

Letters from children

If you have children, you'll have the task of writing – or encouraging them to write – thank you notes for birthday, Christmas and other gifts, plus chatty, newsy letters to doting grandparents and favourite aunts. Here are some tips:

- For young children, you will need to write the letters yourself, although they may like to colour a picture or write their own name if they are able (a handprint or lipsticked kiss is great if not!). Sit down and write the letters with the child, so they get used to the process of writing thank-yous, and ask them for suggestions what to include. Use their words whenever possible.
- If you've several very young children with birthdays around the same date (or at Christmas), it's acceptable to send a letter yourself, rather than one from each child, but be sure to include something from each or a photograph of you all.
- Older children should write their own letters, as neatly as possible, but they don't need to be long.
- For thank-yous to other children it is usually acceptable to use a computer, if you jazz it up with graphics and colour, but the child should always write both names by hand.

- Children have an annoying habit of opening all their presents at once and thus not knowing who gave them what. In this case, you can't do other than say, 'Thank you for my lovely birthday/ Christmas present.' Explaining the confusion can do more harm than good – consider how the reader would react if told that their carefully chosen gift ended up in a heap of other presents.
- If the person who gave the gift wasn't there on the day, encourage the child to tell a little about his/her birthday, the party or favourite presents.
- It is acceptable to use pre-printed stationery, but personalise each message to show that extra appreciation.

Letters to children

Children love to receive letters, so whether you're a gran or grandad, godparent or family friend, don't forget those postcards and 'well done' letters. It's a great way to get them both reading and writing and encourages an interest in what others are doing. Here are some tips:

- If the child has written to you, always thank them for their letter and comment on how well it was written or what it contained.
- For younger children in particular, write in large, well-spaced and well-formed letters so they can read some of the words themselves. Don't cross your sevens, for example, or join your hanging letters.
- Don't use block capitals. Children learn lower case letters first.
- Do use punctuation – young children love to find the full stops and even pre-school children can be taught the idea of sentence structure.

- Young children love 'picture stories' where you replace some of the words with pictures. You don't need to be an artist!
- Use words the child will understand, but don't talk down to them. This particularly applies to children of 8 and above, who hate to be babied.
- Use decorated paper or include pictures.
- Praise their achievements.
- Use humour – a suitable joke, funny story or picture will make the child look forward to the next letter.
- Ask the child to write and tell you about what they are doing, how they are getting on at school/with swimming lessons/on their new computer, etc.

Remember, Beatrix Potter's books started off as letters to a young relative – so there's no limit to what you can do!

Community letters

You might not be the most community-minded person in your neighbourhood, but sooner or later all of us have to get involved. It may be a coffee morning, a sponsored event for your child, encouraging support for a sick neighbour or even summoning help with a difficult issue. You can find examples of such letters on pages 124–127.

Here are some tips:

- Be bright and cheerful when inviting people to events. Use coloured stationery or graphics.
- When raising support or awareness for a community issue, such as the problems of dog mess or a noisy neighbour, don't name names

or use insinuation. Remember that this can constitute libel. State your case clearly and ask for support – a petition slip at the bottom of the letter often works well.

- If you need to ask for funds, assistance or donations, make it clear this is voluntary but point out the benefits to the recipients as well as to those you are inviting.
- Think about the people you are contacting first, and write using words that they will understand and that will encourage them to attend. For example, formal language won't woo teenagers to the Scouts' barbecue, but it's ideal for inviting your elderly neighbours to a genteel garden party on the Scouts' behalf.
- If you're organising an ongoing campaign (such as Neighbourhood Watch), make it clear that only a small amount of time each month will be required and that attendance at meetings, etc, will not be compulsory. The thought of being co-opted on to a committee puts off many people in this situation.
- If you're asking for donations for, say, a jumble sale, make it clear where to bring the goods (offer to collect heavy items and remember to include your telephone number) and add a thank-you.

For further tips on raising funds, see pages 56–57.

Chapter 4

The Awkward Letters

Now, how about those awkward situations where you need to apologise, to tell someone off, to ask for something to be returned – those letters we'd all rather put off? While we can happily harangue that faceless council worker or the inept customer service department, asking a neighbour to stop playing loud music or apologising to a friend for a tactless comment can seem impossible to get right. And that's not all. You may be writing to someone you've only met briefly. You may need to consult an acquaintance in their professional capacity, and not know how to begin. You may have to give bad news, ask for a loaned item to be returned, apologise for a misjudged remark, or change a long-standing arrangement. You may feel duty-bound to offer sympathy or share sadness. All of these can be difficult, no matter how well you know the recipient. Let's look at some examples and tips to make it easier.

Dear neighbour ...

Niggling problems with the neighbours can often turn into a full-blown row if they aren't handled delicately. We're all guilty of this – we don't want to cause a fuss so we shy away from mentioning their loud music, their noisy kids, that overhanging tree or the gutter that leaks over the boundary fence. But sooner or later we have to sort it out. How better than a letter?

- Your first approach should always be informal. Don't head straight for the solicitor or local council.
- Assuming you know the person to say 'hello' to, use their first name. Otherwise, use 'Dear neighbour'.
- Keep your tone friendly, even if you're making a complaint.
- Don't be over critical – but don't apologise if it's not your fault.
- Before you write, think carefully about the circumstances. Does your neighbour play his radio loudly because he's hard of hearing? Are the noisy kids in the garden because they drive their mother too mad in the house? Could you help, rather than criticise?

Examples are given on pages 128–129.

Then there is the other angle – the neighbour to whom we look for advice. Maybe your neighbour is a solicitor, and you need some advice on a potential legal issue; maybe she is a garden designer and you want some ideas for your front yard but can't afford her fees. How do you make that first approach?

- If the matter concerns you both, for example a planning issue that affects both your gardens, your neighbour will probably take the case on for free if you agree to help in some way (telephoning, canvassing support, etc).
- If the matter concerns a third party, ask only for general advice – you can't expect pertinent information for free and your neighbour may in any case be acting for the 'other side'. This would be considered a conflict of professional interest.
- If you need advice but can't afford to pay professional rates, say so up front and ask if they can suggest a good reference book. They

may well offer the advice for free, or suggest another way for you to 'pay', but leave it up to them to choose.

- Offer to discuss the matter over dinner or drinks – you pay, of course.
- Thank them in advance for any assistance they can give.

An example is given on page 130.

Hello, you won't remember me, but ...

These are the letters many of us find most difficult. Let's say you met someone recently at a large gathering who offered to help you with finding a new job. You are wondering whether they really meant it. 'Maybe I shouldn't write ... Maybe I should.' What have you got to lose? Just get on with it!

An example is given on page 176.

- Address the person in the way they were introduced to you. If they gave you a business card, use the form shown on it.
- You don't need to include their address, as this helps to keep it less formal.
- Remind them of who you are, of how and where you met and of how they offered to help.
- Thank them for their time and in anticipation of any assistance.
- If you get no response, don't write again – they are either too busy or have passed on your details but with no luck.

Can I have it back ...?

Sadly, not everyone is particularly conscientious about returning items they've borrowed from you. If it's something small, and the verbal hints haven't worked, you might just want to forget it. But what if it isn't? You might feel awkward about asking, especially if they've told you they'll return it but haven't. Try this approach.

- If someone has something of yours, ask for it outright! Don't hide your request in the middle of a chatty letter.
- A face-to-face or telephone request often works best for that initial attempt.
- Don't feel awkward about asking for your property back – it's yours! If you feel awkward your letter will show it. Don't be hesitant or apologise for wanting it back unless you told them to keep it originally.
- Be friendly – don't threaten. Lighten up the request with a final sentence that proves you're still friends – assuming they return the goods.
- Give a clear deadline, but be reasonable – if someone borrowed something a year ago they may need a few days to find it.
- If your first letter has no response, write another just before your deadline. Don't accuse – remember your reader may have been ill or may simply not have seen the original. Be a little more firm this time and suggest an alternative solution (buying a new item and splitting the cost often works).
- Where money is owed it can be a little tricky. If the sum is large and your letters are having no effect, you may consider legal advice. However, you will need to be able to prove the loan took place and that terms were discussed for its repayment.

For example, see page 131.

I thought I'd better tell you ...

Now how's this for a predicament. You know something about your neighbours' eleven-year-old that you think they ought to know. Maybe he's been bunking off school, smoking weed in the park – or worse! Or you overheard your friend's wife telling her girlfriend that she's leaving him – but you're sure he knows nothing about it. Or you've had a tip-off on the whereabouts of the sound system that was stolen from the local pub on Saturday. What do you do?

First you need to think carefully about how best to tackle the situation. Does it really involve you? In many situations it may be best to just let it lie. But if you feel you can't, consider these points:

- If the wrongdoer is a child, it may be best to approach him or her first and have a stern word or two. Threatening to tell the parents is often as successful as actually doing so – without causing too many ructions at home.
- If a sensitive situation is involved, approach the matter from a different angle. If you're not sure if your friend knows of his wife's intentions, ask how things are and whether they've any holidays planned, etc. His response will help you to see how the land lies. If you still think he needs to know more, it's probably best to do so face-to-face.
- If a crime is involved, then so are criminals, so it is best to let the police handle the matter unless you are sure it can be resolved amicably. Consider the reliability of any information before making accusations. Don't endanger yourself to return someone else's property – a phone call to the police or an anonymous note to the victim will probably do.

If and when you feel a letter is required, write it in plain terms and without accusation or comment on the behaviour. Don't use provocative language and don't include your own role in the matter unless justified. Tell your recipient that they can call you if they want further details or would like to discuss the matter.

Sometimes you may need to write such a letter to a body such as the council, child welfare, social services, etc. We shall look at this in more detail in Chapter 5.

Condolences

What do you say when someone has died? Often you know what you want to express, but just can't find the right way to say it. You need words that will help to soothe and calm in what is often a raw and difficult situation – particularly if the death was unexpected or violent. And you need to be concise – the grieving have little concentration for long epitaphs.

Here are some tips to help you. Example letters are provided on pages 132–133.

- Be yourself – don't try to use formal language if it doesn't come naturally.
- Be respectful, even if you and they don't get on well.
- Write your condolences as soon as possible after you have heard of the death.
- Acknowledge that this is a difficult time for the family.
- Keep it short – no more than four paragraphs.
- Never include your condolences in a letter with other news or information, unless it is a letter to someone not directly connected

(e.g., if your mother has informed you of the death of her neighbour, you can express your sadness to her as part of a more general letter, but should not do so to the bereaved family), or you have already spoken to the bereaved and offered your condolences verbally.

- If you didn't much like the deceased but worked with/for them or you shared positions of responsibility, you are duty-bound to express condolence. Don't be false and pretend to be their best pal – it will be obvious – but instead try to find something noteworthy in their character to comment on pleasantly. A good example is given on page 133.
- If the person died in unusual, embarrassing or questionable circumstances (for example, a suicide, or while with a mistress or as the result of a crime), don't dwell on the manner of their passing but think carefully about what else you say – certain words such as loyal, dedicated or honest might not be appropriate.
- Make it personal: include a happy anecdote about the person who died and their qualities, preferably one that pertains to your relationship with them.
- Certain religions have particular rules for offering condolences and it may be important that these are followed. If the person who died was of a family with particularly strong faith, it may be wise to check with others of the same faith how best to broach the subject.

So sad to hear about ...

The expression of sympathy is not limited to bereavements. People need to feel they are supported in a wide range of situations, from exam failure to marriage breakdown, the discovery of incurable illness and even the death of a pet. Such letters of condolence are rarely easy.

Here are some general tips, in addition to those given for bereavement, above:

- Don't be flippant. No matter how wicked a sense of humour your reader has, sometimes humour is just not appropriate. You may find it funny that Joe's ferret drowned while trying to catch a duck: it's unlikely he does.
- Don't take sides. This is particularly important for relationship breakdown, no matter how final it seems. Couples do get back together, and then those remarks you made about his girlfriend could prove rather embarrassing. You can support without getting vindictive.
- Be sensible. If something is incurable, don't say you hope they get better soon. Don't suggest a new pet can replace the old or that you got where you are today without exam success.
- Try to encourage and sympathise. Don't tell them to get on with it, pull themselves together or forget what's happened.
- Wherever appropriate, offer help or a shoulder to cry on. It is usually appreciated even though the offer is rarely taken up.

An example is given on page 134.

I'm sorry!

Sorry, it seems, really is the hardest word. Many of us seem totally unable to use it, although it's the simplest and often most acceptable way of saying 'It's my fault!', 'I messed up!', 'I got it wrong!'. Officials are particularly good at saying everything that's required, but then they are often paid to do so. If you're not, just get on with it! A simple 'I'm sorry' can often disarm the most rabid critic.

Examples of apology letters are given on page 135.

- Be contrite, not bolshie.
- Don't waffle and make excuses. If you've messed up, whatever reason you give is bound to look like an excuse! If you've received a complaint that asks for explanation, you must always give one, but do so concisely and without placing blame elsewhere.
- Don't overdo it. One 'Sorry' will do – repetition is likely to look like sarcasm.
- Ask for the reader to accept your apology, or to forgive and forget. Then it's up to them.

Please go away ...

> If you are receiving nuisance mail, e-mail, texts or phone calls that are offensive, sexually explicit or threatening, you should immediately contact the police. Do not destroy or delete any such item until you have clearance from them to do so.

We all have the problem of junk mail – which we will look at later – but there is another type of nuisance mail that is more personal and causes more anxiety – that from people you don't want to hear from: the unwanted admirer; the long-lost friend who won't take no for an answer; the scrounging relative or the couple who invite you to stay when you'd really rather not. In many cases the best option is first to decline, then if the requests don't stop, to ignore them and hope they give up. But some people just won't. This can be difficult, and may result in legal action, so it is best to deal with it in writing and in a formal manner. Consult a solicitor if necessary, but often one or more polite but plain-speaking letters will be sufficient to stop the flow. An example is given on page 139.

If you are being bombarded with e-mail from a known source, it can be relatively easy to stop this: most e-mail software allows you to 'block' or 'screen' incoming mail from a specific e-mail address. This doesn't work well with genuine junk e-mails, known as 'spam', such as that from companies, as they frequently change the sending address to get around such blocks. But it can work when a particular person is sending nuisance messages. Check your e-mail software for details.

Then there is the annoyance of being added to a friend's mailing list for e-mailed jokes, anecdotes and images (often explicit). In this case, the annoying mail is often only part of what you receive from this person, and you need to keep the rest (for example, if you are receiving mass-mailed jokes from someone at work). A simple rebuke is often enough here, particularly if you hint at repercussions (see the example e-mail on page 140).

Nuisance texts can be difficult to stop. Some phone providers now allow you to block specific numbers, as you can with voice calls (ask your own provider for details), but often the best solution, if you are receiving a lot of nuisance or offensive messages from a particular source, is just to change your phone number, giving the new number only to those you trust. Again, the police can help if you have evidence of continued harassment by text. Replying (or preferably sending a letter) with an indication of your intent to contact the poice will often deter offenders from continuing. Remember:

- Ignoring the correspondence might work – try it first.

- If this doesn't work, try returning letters, marking them 'no longer at this address'.
- If the sender still doesn't stop, take a firm stance from your first reply.
- Keep copies of all correspondence received from and sent to the offender.
- No matter how annoying the problem, be polite when asking for it to stop.
- If the mail becomes offensive or threatening, take copies to the police.

Appeals for donations or bequests

These days we all receive so many begging letters in the form of circulars and charity requests that it can be very difficult to get the attention of your reader when the Scouts need new canoeing equipment or the school roof needs repair. Even more emotive pleas for support for a local hospice or to raise funds for the children's ward can be binned with the junk mail. So how do you make that appeal really register and get your readers' hands in their pocket?

- First, consider whether there is an alternative way of raising funds that may be more appealing without requiring much investment – a race night; a coffee morning; a jumble sale. Often a small entry fee can be levied and a raffle held which will help raise funds, and a request for loose change on exit can reap quite substantial rewards.
- If this is not appropriate, you may need to write a letter directly appealing for donations. To help with this, collect up some of the commercial begging letters you receive over several days. Consider which ones work and which don't. Make some notes on the things that grab your attention and those that really turn you off.

- To be successful, your letter must appeal directly to the reader. While there are a few genuinely benevolent folks out there, most people are by nature self-interested. Therefore if you're looking for donations for a home for sick children you'd be more likely to get a response from couples with children – especially those whose children have been in hospital in the past – than from a group of single twenty-somethings.
- Be sincere. If you have a personal reason for asking for support, mention it. If you explain your involvement, your neighbours are more likely to feel obliged to donate than if you don't.
- Explain the benefits of such a donation to both the donor and the recipients. Mention any applicable tax benefits – the thought of grabbing a few pence back from the taxman softens the hardest heart.
- He who pays the piper calls the tune. Aim for those to whom you've donated in the past – they should feel obliged to return the favour.
- Keep your letter simple. While many causes are emotive, use of too many adjectives describing suffering and need can be off-putting. Remember that the best way to get through is to be personal.
- Wherever possible, write your requests by hand – it will help them stand out from the rest and will make them more personal.
- If you are creating brochures or supporting information, don't make them too flash. Critics will say that you've wasted money that could be going to your cause. On the other hand, you need to look professional when appealing to those you don't know personally or they may think it is a con.
- If you want your appeals to be successful, don't do them too often or for more than one or two recipients. Repeated requests for a variety of charitable organisations will result in your letter being consigned to the bin along with the rest of the junk mail.

Chapter 5

Getting Things Done with a Letter

No matter how we try to avoid it, eventually we all need to write a letter to someone we don't know well. It might be a business or local authority, a person doing work for us, or someone else such as a teacher, doctor or bank manager. Many of these people will receive piles of letters each day, often on the same subject. So how do you get yours noticed and make sure it gets you the answers you want?

Many of us find this difficult, so don't worry if you're not sure where to begin. First, read the sections in Chapter 1 about how best to get your questions and points across to achieve the result you want . Then return to this chapter where we'll look at the things you need to consider for particular situations and how to make sure your letter achieves its aim.

Letters to the council or local authority

Council tax, planning permission, the state of public toilets, rubbish collections. All of these affect us at some time and they are all dealt with by that amorphous body, the local council. I don't know about you, but there are almost as many letters to my local council in my letters file as there are to my bank. So what's the best way to get your letter noticed?

- First, find out which department deals with the problem or the subject you need to discuss. The best way to do this is simply to phone the local council's switchboard or look at its web site.
- If possible, get the name of the person responsible. This will help enormously in tracking the progress of your enquiry.
- Write in a formal style, addressing your letter directly to the person concerned, and quoting all relevant references (such as your council tax reference, or the exact location of the problem you are reporting).
- One issue = one letter. Don't complain about the litter from the local takeaway in the same letter as an objection to the extension on your local supermarket.
- When complaining, don't whinge. State the facts baldly and say how they affect you and others. Propose a solution, or ask your reader to suggest how you or the community might help to resolve the problem.
- Thank the council official for his or her time and efforts in resolving the situation. This really does go a long way.
- When the problem is resolved, if you have been happy with the way the council has handled it, write to your contact and say so. Hopefully then you will be remembered and next time you have a query it will be met with a positive response.
- If you cannot get the problem resolved to your satisfaction, ask whether there is an ombudsman or another authority to whom you can appeal. If it is a matter that concerns your neighbourhood, a petition often works well.
- When writing to a specific councillor, rather than a department, use the form of address given on page 110 and consider the points shown below for letters to politicians.

Some examples of letters to the council are given on pages 143–144.

Letters to your MP and government bodies

The role of all MPs is to support their constituents and represent their views, so make use of it!

Write to express your dissatisfaction at a proposed change in the law or government policy; to express your support for a campaign; to request that a matter of personal or community importance is raised in the House; even to congratulate him or her or the party on a particular policy or achievement.

You'll often find persistence is required when asking MPs to support you on an issue, simply because they receive so many spurious complaints about small problems. But if you write your letter well, stating the problem up front and explaining how they could help, it makes a big difference. The same applies to letters to civil servants working in government departments and semi-government bodies. Here are some tips; see pages 144–146 for examples.

- First, find out who your MP is. If you don't know, look it up on www.askyourmp.co.uk or ask at your local library.
- Find out all you can on the issue that concerns you and create a list of the reference material or contacts available. This will allow your MP or his or her staff to research the issue more quickly and will help to keep your letter shorter – making it more likely to be read.
- Use the correct form of address. It shows you have taken the time to do it properly. See page 109 for details.
- Include a reference line showing the subject of your message, and outline the problem in the first sentence.
- Try to keep the letter to one page, two at the most. You can include supplementary material (see above) in addition to this so you can reduce the length of the letter by sticking to the main points.

- If the matter concerns your community specifically, point out that it is your MP's responsibility to act in the interests of constituents and that you are requesting help on this basis.
- If the matter is more general, but you wish to register your support or opposition (say, for a bill tabled to be read in Parliament), make this clear. Ask for the points you raise to be put to the proposer of the bill and request answers.
- Your MP may reply saying that he or she cannot deal personally with the matter but has passed it on to a particular government department. Make sure you get contact details and then forward both your original letter and a copy of the MP's to that person, asking for comments as soon as possible. This two-pronged attack often gets better results.
- Often an MP will reply stating that they sympathise but cannot vote against their party's line. This is a classic defence tactic. Reply stating that in any case you wish your opposition/complaint to be heard and that as your MP he or she is responsible for making sure that your views are represented.

Letters to schools and the education authority

If you have children, it's likely that you'll eventually have to do battle with your local education authority, ask the school for time off or make a problem known. With many functions now being taken out of the hands of individual schools, who know their pupils, and handed over to faceless centralised departments, getting your message across can be difficult.

We've some successful examples on pages 152–155. Think about the following when writing your own letter:

- Emotive language rarely works when trying to convince someone to do something on your behalf. Stick to facts.
- Consult with other parents to muster additional support.
- When asking to take your child out of school during term time, explain why. Always ask, don't *tell* the school he or she will be absent.
- When writing to excuse your child's absence because of illness or other valid reason, keep it short and give a reason why you were unable to advise the school in advance or at the time.
- If writing to complain about another child's behaviour, use calm language and don't accuse unless you have the evidence to prove it. Remember the letter may be shown to the child or its parents.
- When disputing a result, decision or ruling, quote any reference number and refer to the chain of events that led to this point. Acknowledge the school/council's input into the situation and explain how you believe their decision to be wrong. Supply any additional facts or evidence with your letter.
- If you don't understand their reasoning or the process that has been followed, ask – they are legally obliged to explain it to you.
- Don't expect miracles. Some children will always have to go to school some distance away, just as some will always be bullies or bullied. Schools never have enough places, books or equipment and there will always be a few disappointed parents. But persevere until there is no avenue left.
- Don't blame the school for policies handed down by government or local education authorities – in many cases it is out of their hands and they are unable to intervene on your behalf. Try to keep on good terms with the school in all cases.

- Remember, the better the school knows you, and the more you do for them, the more likely you are to get your points listened to and considered, even if you don't always get the result you want. So get involved in support groups and classroom assistance, show an interest in your child's progress and help out when asked. Make sure your name is known by those in charge!

Letters to the media

Writing to complain about TV programmes or the redesign of the Sunday supplement is not merely the province of retirees; many of us get incensed about the reporting of particular issues or the (non) coverage of events. Most of us just grumble and do nothing about it. But if you want to make your opinions heard, there is often an avenue for you to do so.

- Consider the best option for your complaint or suggestion and write to the appropriate authority. Some useful addresses and programme details are given on pages 186–187.
- In the subject line of your letter, name the programme/episode/article you are writing about, including the date (and time, where more than one show is presented each day). Include the page number for printed items.
- As with all letters of complaint, don't whinge and don't get personal. State the facts calmly and then outline your problem. Ask for the reader's opinion and where appropriate for details of the action they intend to take to remedy the situation.
- If you are writing to ask for a copy of a programme or transcript, offer to pay the postage and any other associated charges. Indicate

the format you require (e.g., VHS, DVD, word-processed file or paper copy).

- Thank your reader in the closing paragraphs after summarising in a single sentence what you have already covered.

Some examples can be found on pages 155–157.

Letters to shops and businesses

Although much more commerce is now taking place in person or via the internet, there are still occasions when you will need to write to a business. You may have ordered goods that never arrived, or which were damaged on receipt; you may need to correct a transaction or ask for a refund. You may wish to protest at a proposed expansion or the closure of your local branch. Or you may need to complain about the service you received at a hotel or restaurant.

- All letters to businesses should be written in a formal style, no matter how small or informal the business. Include both your and their full address, and any reference numbers or order codes.
- Where possible address your letter to a specific person. Often a quick telephone call will help establish the best person to deal with your query. If not, address it to the Customer Services Manager, or the Managing Director if a small firm.
- If complaining about goods or services, get your facts straight first. Make sure you have copies of receipts, order confirmations, all relevant dates and – where possible – names of those you have dealt with. If complaining about the service received in a restaurant or hotel for example, include the time of the booking, the table or

room number, the names of the staff concerned (if you got them) and a copy of the receipt.

- In the subject line, state the product name, model or serial number or any invoice or order reference.
- Think about what you want. The goods you ordered? A refund or compensation? A replacement product? An apology? For the product to be removed from the shelves? Make this clear from the outset.
- Keep your letter as short as possible, while stating all the facts that pertain to your case. Enclose copies of all relevant documents and list them under 'Enclosures' at the end of your letter.
- Ask the reader to acknowledge receipt of your letter. This way you know that it has arrived and your reader is more likely to deal with it immediately.
- If you receive a bill for a product or service that you did not receive or order, for goods you returned, or that is incorrectly calculated, respond as quickly as possible to ensure the error is corrected. Do not be overly critical in your letter; everyone makes mistakes!

Some sample letters are given on pages 158–159.

Letters to banks and other financial institutions

When writing to any financial institution, be it a bank, building society, insurance company or credit card company, you need to be especially clear in your instructions and requests. Although more and more of us are using internet or telephone banking, there are still occasions on which only a letter will suffice, particularly when opening or closing an account or arranging complicated transfers; and the great benefit of banking by letter is, of course, that you can retain a copy.

- Always quote all relevant policy or account numbers and the names in which they are held. Check your chequebook or policy documents for these as it is important that they are correct.
- With bank payments or transfers, also quote the sort code and the name in which the account is held, and at which branch, and clearly state on which date the transfer is to take place.
- For international transfers make clear the currency to be used and to which party any commission or fees are to be charged. Give either the date when the money should leave your account or the date on which it must arrive; sometimes this can take a week or more so it is important to get this correct.
- Be honest. If you are having difficulty making payments on a loan, for example, state why and when you expect this to improve. Suggest a compromise that the bank is likely to accept (for example, smaller payments for three months, then stepped increases until you are back on your feet).
- Letters to banks must always be signed in the same form as you use on your cheques and credit card vouchers. Instructions can generally only be accepted when signed by the person holding the account (or an authorised signatory in the case of companies); joint signatures may be required for some jointly held accounts.
- When requesting a loan or overdraft, write to the manager of your branch or the person responsible for maintaining your account.
- Mark all letters regarding your account as 'Private and Confidential'.
- Always retain a copy. This can be a carbon, a photocopy or a computer file, as long as it is a true replica of the letter you sent.

For examples see pages 160–161.

Letters to your landlord or tenant

At some point many of us live in either rented or leased accommodation, where repairs and structural alterations are the responsibility of another person. Whether you pay weekly rent or an annual maintenance charge, you can use the same approach when you need to get something done or when you need to change the way you pay. Some of us may also rent property to others – your empty house while you're working away, or simply the spare room. Inevitably there will be situations where you need to write to your 'tenant' – although a face-to-face discussion first will usually resolve the situation.

When dealing with landlords or tenants:

- Be aware of your rights. If you have a tenancy agreement, re-read it thoroughly before making your complaint or request. If necessary, seek additional help from your local housing advice centre.
- If asking for repairs, make the landlord aware of the need for the repair as soon as possible. If nothing is done, follow up with a request for the work to be done and a completion date.
- If the repair or problem affects more than one tenant or property owned by the landlord, seek the help of the others in resolving the problem. Often a letter signed by you all will get better results than an individual moan.
- Don't nit-pick with your tenants. This can be a temptation especially if they are sharing your house, but not everybody may have your impeccable hygiene and tidy nature. If their room is a mess, don't complain, just shut the door. If the mess finds its way into communal areas, however, you do have the right to tackle them about it, but only in extreme cases will you need to write. If, however, they are

leaving items about that could be a danger to your health, or that are unsuitable material for other members of the household, a letter is often the best way to ensure the behaviour is not repeated.

- Be polite and be patient. Your landlord – or tenant – has other responsibilities as well as your property. Harassing him or her will not necessarily get it resolved more quickly.
- If you dispute a rent increase, check first with your local housing advice centre or seek opinions from local letting agents on whether the sum asked reflects the market rate. If not, ask for written estimates or reasons why there is a difference, and submit these with your letter to your landlord, asking him to explain or reconsider.
- If you are asking your tenant to pay more rent, or overdue rent, explain why and offer to discuss the matter if this causes financial difficulties. You may be able to come to some other arrangement – say with the tenant performing set household chores or doing the gardening – in lieu of the increase. Put this arrangement in writing and ask the tenant to sign it.
- If you are asking for more time to pay your rent, do so as soon as possible. Explain why you are unable to pay and apologise for the problems this may cause. Try to suggest other ways in which you can help to make up for this (see above).
- If you are writing to terminate your tenancy, do so as soon as you know but give a definite moving-out date to avoid any confusion. Arrange a date and time at which the property can be inspected in order that any deposit can be returned, and supply a forwarding address for any mail.
- If you are writing to terminate your tenant's tenancy, give them as much notice as possible and explain why they cannot remain in the

property. If they have breached the tenancy contract, say, by not paying the rent, there may be legal consequences, so you should consult your property advisor or solicitor before writing.

Examples are shown on pages 162–168.

Letters to your employer

Many employment issues can be dealt with by phone or in person, but there are occasions when only a letter will suffice. These include resigning, requesting a promotion or transfer, or asking for facilities to be provided or complaining about your work environment.

Some examples are given on pages 169–174.

- Think first about who is best placed to deal with your letter. If it concerns the canteen food, for example, your line manager may not be able to help but the catering manager will. If it concerns your career path, it may be best directed at the personnel manager, or your immediate superior, depending on the structure of the company.
- Always write to a named individual and state your position in the company and, where necessary, the location in which you work.
- If the letter concerns you personally, write using your home address and mark the letter 'Private and confidential'. If you are writing on behalf of colleagues or a department, you can use your office address.
- If you have spoken to this person before on the subject, refer to the conversation by date. If you have received previous correspondence on the matter, say so and quote any reference given.
- Retain a professional tone, remembering that your letter may be seen by other employees. Do not use libellous statements or slang.
- If you are making a request or complaint, thank your reader for taking the time to consider it and offer to discuss it face to face.

Job applications

There is a certain etiquette about job applications that must be followed if you are to succeed in obtaining that new position. In most cases, a well-styled and detailed curriculum vitae will be required, and you should read the various publications listed on pages 183–185 to help to maximise the potential of your own qualifications and experience in such a document. A CV cannot be sent alone, however, so you will always need a covering letter – an example can be found on page 176.

Some job applications do not require a CV. Here your covering letter will need to be a little more detailed to explain your interest and qualifications for the job. An example is provided on page 177.

When writing job applications:

- Remember that your letter is your ambassador. It will create an impression of you in the mind of the reader, and we all know that first impressions count. You may need to write several drafts before settling on a final version; present it on quality paper with a suitable typeface or in neat handwriting. Don't use coloured paper or print.
- If you are writing in response to an advertised position, address your letter to the person named in the advertisement. If not, address it to the Personnel Manager or to the Managing Director if it is a small firm.
- Think about what the person reading your letter most wants to know. Often one of the main considerations, apart from relevant skills, is your availability – can you start immediately? If so, say so! If not, indicate the amount of notice you will be required to serve with your current employer.

- If the job advertisement requests specific skills or qualifications, state whether you have these, and if not, how you intend to gain them and why you think you are suitable for the job without them.
- Never overstate or lie about your experience or qualifications. Indicate your willingness to undergo training, where appropriate.
- If you are writing to ask a firm to consider you when a job has not been advertised, you need to sell yourself and your skills even more. Speak of your abilities, any qualifications and experience, and suggest how you think these would best benefit the company. Ask for a meeting with the company to discuss a suitable position and ask for your details to be retained for any future vacancies. You may need to write several of these letters to a number of employers before you get a response.
- If you receive a notice of interview or a job offer in writing, always respond to it in writing.
- If you receive a letter stating that you have not succeeded in obtaining the job, and you would still like to work for the organisation, it does no harm to write thanking the interviewer for their time and asking for your details to be kept on file for further vacancies. You may also ask for feedback on your interview performance or the reason for your rejection.

Letters of reference

There are several situations in which you may be required to provide a character reference. These include:

- Applying for a new job.
- Applying for a mortgage or other loan or finance.
- Applying to join or progress within a club or community group.

The requirements for each differ. For example:

- Employers are primarily interested in how well you could do the job you have applied for, and whether you'll fit in to their organisation. But they also need to ensure you are honest and trustworthy. They'll want this information from someone you've worked with or for, or someone who's been directly involved in your education, if it's your first job.
- If you are applying for a mortgage, you may be asked to provide a reference from your employer to prove your income, the likelihood of continued employment, and any prospects for promotion.
- If asked to provide a reference in order to join a club, it is likely that the requester is less interested in your financial position (unless dire) than in your overall honesty and integrity, and a confirmation that you possess any required skill or experience that you claim to have.
- To take on some community responsibility such as Neighbourhood Watch co-ordinator, a reference is often mandatory. This is to prove that you are of suitable character – which often means being a good communicator, even-tempered and willing to listen, as well as honest and reliable – and have no relevant police record.

Asking someone to provide a reference

Think carefully when selecting someone to provide a reference, making sure they are equipped with the relevant facts and figures to provide the information required. If asked to give a reference when applying for a new job, you may prefer to ask a previous employer rather than reveal your job search to your current employer. An example of how to do so is provided on page 181.

Always write (rather than phone) to ask the person to provide your reference, so you can show the bank, employer or club you have requested it.

Try this approach:

- First, remind the person who you are. If you work for a large organisation, they might not remember you immediately or fit the name to your face.
- Explain why you need the reference, and why you've asked them.
- Thank them for their assistance
- Provide contact details.

Giving a reference for someone else

You may also be asked to give a reference. Consider carefully:

- Have you known the person well for some time?
- Can you provide all the information requested?
- Can you honestly attest to the person's character or skill level?

If the answer to any of these is no, someone else may be better placed to provide the reference. You should explain this to the applicant as tactfully as possible – often it is best to do so face to face or by telephone rather than a letter. Suggest someone more suitable if possible.

When asked to provide a reference, first find out whether it is for a particular job or request, or whether it is to be used for several applications. School leavers, for example, often request an all-purpose reference that they can send out to several employers, and these require a slightly different approach.

Let's look at some tips for writing specific references:

- Before trying to provide a reference, talk to the applicant to find out why it is required and what the person requesting it (employer, club secretary, etc) is likely to want to know. Find out whether there are any particular aspects the applicant wants you to stress.
- If you're not sure what the person requesting the reference actually needs to know, contact them first and ask. Don't get drawn into providing a reference by telephone; it is something that needs to be considered in depth before putting pen to paper.
- If the reference is for a job, find out what the job is, and consider whether the applicant is really suited to it. If it's for a club or social purpose, think about the person's team spirit and temperament, as well as whether they have the time to dedicate to any responsibilities involved.
- Write directly to the person requesting the reference, and mark the letter 'Private and Confidential'.
- For work references, explain your position in relation to the applicant's; how long they have worked for the company; what their job entails and how their skills are deployed; what training they have

had; how well they work and interact with others; any social or leadership responsibilities; and how they have progressed.

- For social references (e.g., for a club membership application) provide information on the applicant's ability to relate to others; sociability; honesty and integrity; social conscience (any prejudices, for example); and, where appropriate, their skill or achievements.
- For community-related references (e.g., for a person applying to be the Neighbourhood Watch co-ordinator), comment on the applicant's ability to get on with others; to motivate and enthuse; their ability to allow time for the work associated with such a responsibility; their honesty, integrity and reputation.
- Don't overstretch yourself. If you know the person only in a work environment, don't try to comment on their social life, and vice versa. Make it clear in the letter that this is the case. If possible, provide the name of another person who may be able to comment more accurately on that aspect of the applicant's life.
- If you feel the person would be really unsuitable, you must say so, but do stress their skills and positive qualities.
- Give a contact number in case they require further information.

An example is given on page 182.

Chapter 6

E-mail – The Speedy Alternative

E-mail has been hailed the saviour of letter writing in an age when few of us have the time to put pen to paper regularly. One message can be sent to many readers at the click of a button, so it is ideal for chatty family messages, party invitations or organising meetings. But is it as suitable for communication with your bank, your landlord or your plumber? And anyway, how do you use it?

> Don't have a computer? You can still e-mail. Try your local library, advice centre, adult learning centre or coffee bar. Many are now equipped with internet facilities that you can use for a small fee, and most even offer help with getting started. Don't be shy – get online!

Is e-mail the new letter?

E-mail is used widely in business, and in many cases has come to replace the typed letter or memo. But not in all. For example, while many jobs are advertised on the internet and may be discussed via e-mail, it is not considered appropriate for hiring (or firing) to take place via e-mail. While goods may be ordered via e-mail and initial comments or complaints made in the same way, any protracted correspondence should revert to paper. However, most inter-office communication these days is made by e-mail and it is certainly ideal for disseminating information to a large group.

So how about personal messages? If all your family or friends are online, is e-mail acceptable for thank-yous, happy birthdays and the Christmas 'round robin'? Why not! It has great benefits:

- You can send the same news to many people at once.
- You can include scanned photographs, links to web sites, video or audio clips.
- You don't have to remember lots of addresses.
- You can send e-mail from anywhere in the world.
- Short messages are as acceptable as long, newsy letters.
- It's much quicker than the post, even to far-flung destinations.

However, as it is considered a fairly informal medium, e-mail should never be used in sensitive situations, when a letter, preferably handwritten, is by far the best solution.

How about those matters that fall between 'personal' and 'business' though? Can you use it to book a table at your favourite restaurant? To ask your landlord to fix a leaking pipe? Here you have to use your discretion. Remember that:

- Not everybody with an e-mail address reads their mail daily.
- Electronic communication is no more reliable than the post – and usually you can't check whether a message has arrived or not.
- E-mail is easy to delete or ignore – especially if your recipient has many messages arriving in one day.
- E-mail is informal – a letter often has more impact.

So I'd suggest you revert to the telephone for that restaurant table, but if the problem is less urgent, e-mail will probably be fine. You can always follow it up the next day if you've had no response.

E-mail comes into its own if the situation needs some lengthy explanation or figures are involved. For example, if you want to ask your bank for initial approval on a loan. Here a quick phone call to brief the bank manager on your outline requirements could be followed up with an e-mail detailing your income, the proposed purchase, and other relevant information. This will be appreciated by the bank as it saves the need for you to make an appointment, and the need for them to write it all down during your phone call. They can transpose your figures directly into their own records and give you an answer far quicker.

The practicalities – getting yourself online

These days, you don't need your own computer to e-mail, although it helps. You can also e-mail from:

- Libraries and advice centres.
- Some coffee shops (often called 'cybercafés').
- Schools and adult education centres.
- Offices (if your employer allows it).
- Some mobile phones (if WAP-enabled).
- Some telephone kiosks, particularly at airports and major rail stations.

The advantage of these is that you don't need to have and pay for your own telephone connection to the internet. You may have to pay a separate fee, however, and e-mailing from mobiles can be particularly costly. At the time of writing, e-mailing from any type of phone is fairly basic and does not permit the sending or viewing of any content other than basic text, but this will no doubt change as technology develops.

If you have your own computer but no internet access you will need the following to be able to send e-mail:

- A telephone socket close to your PC.
- A modem or broadband connection device.
- Some software to allow you to connect to an internet server.

Most new computers come with modem and software installed, but if in doubt ask your local computer shop for advice.

Connection software is often distributed free, via the post or on the counter at computer retailers or newsagents. However, in most countries you do still have to pay for the cost of the call or connection, and this can vary widely. So don't just plump for the first software that you see: check out several and think about how you are going to use the internet first, before you sign up to something that could work out rather expensive. The options are usually:

- Pay a fixed fee per month for a fixed amount of time you can spend online.
- Pay a higher fixed fee per month for unlimited time online.
- Pay no fixed fee but pay for each call you make to the internet.

Which you choose will depend on:

- How often you intend to use the internet and for how long.
- What time of day you intend to use the internet.

- Whether the number you call to connect to the internet is charged at local rate.

If you have cable television and telephone you may be able to get a special deal via your cable operator. This will usually require you to use a cable modem rather than the one installed in your computer. Check with your cable operator for details.

Broadband access gives a faster, higher quality response from the internet than the usual dial-up connection, but is still quite expensive in most countries, often involving a set-up fee plus a monthly charge for a minimum term of one or more years. For e-mail it offers no major advantage, so if your internet activity is going to be limited to sending messages you don't need it. It's probably best to start out with a pay-as-you-go dial-up connection for the first few months to see how you get on before committing to a long-term deal.

What does e-mail look like?

> As all systems differ, we can't show you how to install and use all the different types of e-mail software, or how to log in, open, send and receive messages. Instead we suggest you read the instructions supplied with your connection or e-mail software, or use the online help (usually available from a Help menu).

The appearance of e-mail depends upon which e-mail software you are using, but a few typical examples are shown over the next few pages. As you can see, it is quite simple and informal, with no need to include addresses or other elements that you would expect in a letter.

Instead of a reference line you put any pertinent reference in the 'Subject' line, and the date and time is added automatically when you send the message. For this reason most e-mail messages comprise only:

- The subject line: 'Plumbing problem at No. 14'.
- A greeting: 'Hi Susan' or 'Dear Mr Brown'.
- Main text: short paragraphs!
- A friendly close: 'Regards' or 'Best wishes'.
- Your name (and address/telephone contact if required).

For example:

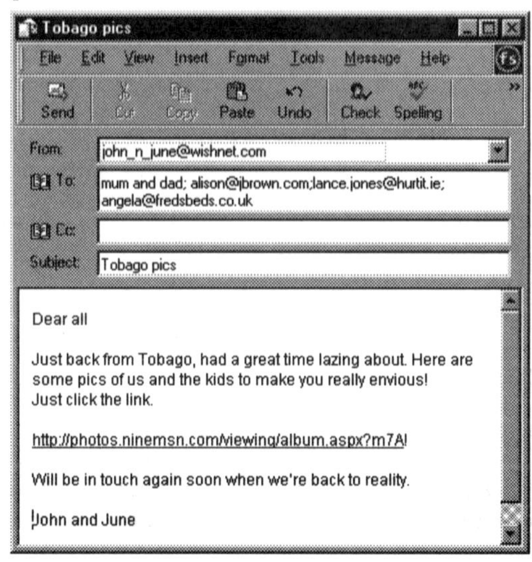

With many e-mail programs you have the choice of writing in plain text or in HTML, sometimes called 'rich text'. (This is usually an option on the Format menu.) HTML allows you to include colour, different typefaces, backgrounds and bullets within your message, whereas plain text is just what it says – plain text. Not all e-mail software can interpret the HTML extras, so plain text is usually better for that first approach. However, if you receive a message from someone in HTML it is usually OK to reply in the same way.

Tips for better e-mail

- **Keep it short:** If you need to provide a lot of detail, write it into another document and attach it to the e-mail message (see below), which can then be a simple covering note.
- **Keep it friendly:** Formality is not required but remember that your message should still be polite and that abrupt remarks can be taken the wrong way. Always read through your message critically before you send it and correct any ambiguities.
- **Keep it simple:** Don't bother with fancy fonts and colours, background 'stationery' or graphics unless you are sending invitations or fun items to friends. As discussed above, they don't always appear to your reader as you might expect.
- **Keep it safe:** Never put any confidential data into an e-mail, particularly credit or debit card information, PIN numbers or passwords, or anything that could be used by someone else to access your money, your home or your personal information.
- **Keep it small:** If you are sending pictures or other attachments (see below) use a compression program such as WinZip (available free from www.microsoft.com) to pack them first. This helps to speed

the message along and reduces the chance of it becoming lost or corrupted or being read by someone else.

- **Observe netiquette:** The unwritten code of e-mail communication. Don't use block capitals (called SHOUTING); always give a measured response (don't 'flame'); don't try to be funny or sarcastic (see below); always complete the subject line of your message; and always check the address is correct before you send your message. For more details on netiquette, see the sites listed on page 185.

A warning!

Because e-mail messages are written informally, as though you are speaking, but have no associated tone of voice or visual clues for the reader to pick up, they can often appear to be rather abrupt – even offensive – when that is not the intention. Humour does not come across well and remarks made innocently and hastily may be read with completely the reverse meaning. With practice and experience, you'll make fewer mistakes in both the writing and the interpreting of e-mail, but do beware, particularly with those you don't know.

To avoid offence, we suggest you:

- Avoid humour and clearly identify any jokes with a smiley.
- Do not use sarcasm or tongue-in-cheek criticism unless clearly identified before and after.
- Use appropriate emoticons (see page 97) if your reader will understand them.
- Be polite and courteous. Don't forget those ps and qs.
- Don't forward on any jokes or similar material that you receive unless you are sure it will not offend the person you are sending it to – or their family or colleagues, who may well read it first.

Getting the tone right

In view of the above, here's an example of how to – and how not to – get it right with e-mail. The first example is bright and breezy, but also dreadfully long-winded, ineffective and somewhat irritating!

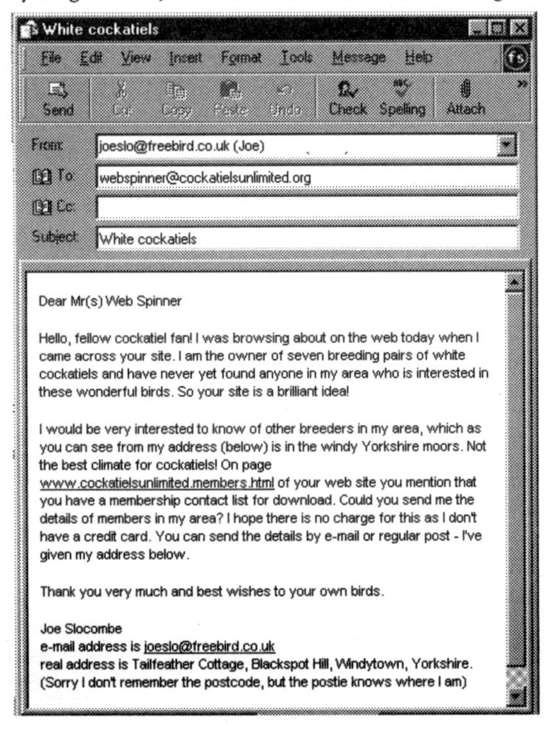

White cockatiels

From: joeslo@freebird.co.uk (Joe)
To: webspinner@cockatielsunlimited.org
Cc:
Subject: White cockatiels

Dear Mr(s) Web Spinner

Hello, fellow cockatiel fan! I was browsing about on the web today when I came across your site. I am the owner of seven breeding pairs of white cockatiels and have never yet found anyone in my area who is interested in these wonderful birds. So your site is a brilliant idea!

I would be very interested to know of other breeders in my area, which as you can see from my address (below) is in the windy Yorkshire moors. Not the best climate for cockatiels! On page www.cockatielsunlimited.members.html of your web site you mention that you have a membership contact list for download. Could you send me the details of members in my area? I hope there is no charge for this as I don't have a credit card. You can send the details by e-mail or regular post - I've given my address below.

Thank you very much and best wishes to your own birds.

Joe Slocombe
e-mail address is joeslo@freebird.co.uk
real address is Tailfeather Cottage, Blackspot Hill, Windytown, Yorkshire.
(Sorry I don't remember the postcode, but the postie knows where I am)

Try this instead: simple and polite, it explains all you want to know, and you're less likely to be dismissed as a crank!

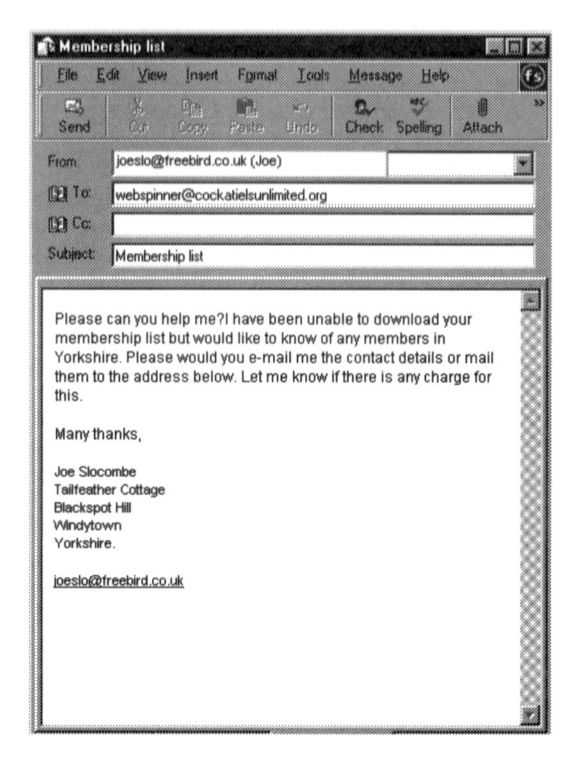

Membership list

File Edit View Insert Format Tools Message Help

Send Cut Copy Paste Undo Check Spelling Attach

From: joeslo@freebird.co.uk (Joe)

To: webspinner@cockatielsunlimited.org

Cc:

Subject: Membership list

Please can you help me?I have been unable to download your membership list but would like to know of any members in Yorkshire. Please would you e-mail me the contact details or mail them to the address below. Let me know if there is any charge for this.

Many thanks,

Joe Slocombe
Tailfeather Cottage
Blackspot Hill
Windytown
Yorkshire.

joeslo@freebird.co.uk

Writing and sending a basic e-mail message

1 First you will need the e-mail address of the person you want to write to. If it's someone you know personally, the best way is to ask them to send you an e-mail (which will then automatically show their address). If not, you may need to look up the address on their business card or even their web site.
2 When you have the address, type it into the 'To:' line of the message. Most e-mail software also allows you to add the address to a book, so you don't have to type it out again next time.
3 If you want to send copies to other people, add their e-mail addresses to the 'To' or 'cc' lines. (Alternatively, if you want to send messages to a lot of people but don't want them to know who else received a copy, put all the names on the 'Bcc' line and put your own address in the 'To' line.)
4 Give your e-mail a title in the 'Subject' line.
5 In the main text box, type the body of your message. Start with a greeting on the first line, then a line space, then the message, another line space, then your closing line. Remember the rules discussed above.
6 If you want the person to contact you by another means, add the details at the end of the message (for example, your phone number).
7 If you want to add any attachments, click the appropriate button and select the files you want to send (more about this below).
8 Click on the 'Send' button either to send your message immediately or place it in an 'outbox' waiting for you to send later.

Creating and sending pictures and documents

To send text files, documents, photos or graphics of any kind, you can simply attach an electronic copy to an e-mail. For photos, some high-street photo-processors supply images on CD-ROM, or you can scan in the image using a scanner. File the images on the computer. Once you have written the e-mail, press the Attachments button and choose the relevant document.

If your photo-processor offers electronic or digital copies of your photographs, it may be possible for you and others to link directly to their web site to view the pictures, rather than cluttering up your mailbox with lots of copies that take a long time to download. In this case, obtain the details from the company, including any necessary password, and e-mail these to your family and friends.

Don't restrict yourself to photos though: with e-mail you can attach computer graphics, scanned images of documents, almost anything.

Sending sound and video files

Now this sounds complicated. But it isn't! If you have a digital camcorder you can transfer video directly to your computer (see your instruction manual). If not, and you have a web cam as part of your computer package, there are ways to record yourselves and you can send the resulting video file as an e-mail attachment. As with photos (see above), some companies will process your video and place the results on their own secure web site for you to access, and you can e-mail this link to your family and friends. You – and they – may need to install some basic software first, but this can usually be done automatically when you try to open the file, assuming you have sufficient space to store the files.

If you have a microphone or headset for your computer you can use very basic audio software (Microsoft Sound Recorder – which comes as part of the basic Windows package) to record your voice and send it as an e-mail attachment or as the background (in Outlook Express – from the Format menu of the message window) if you want it to play automatically. It's ideal for giving really personal greetings if you can't be there in person.

You can also e-mail the latest sounds from web sites, or even clips from a CD-ROM, as attachments or backgrounds too. Just make sure you're not infringing anyone's copyright when you do so.

Who's reading my e-mail?

Because e-mail is used so much in business these days, it is possible to make it far more secure than rumour suggests – but often only with the purchase of additional security software. The prevalence of irritating and destructive viruses or worms makes it essential that you install suitable anti-virus software to protect both your own computer and those of anyone you correspond with.

For most messages you need have little fear, but be sensible:

- Don't send your personal bank/credit card details by e-mail. Internet shopping sites use special encrypted servers for credit transactions – but your general e-mail doesn't pass through one of these.
- Don't send passwords or login details, codes for your burglar alarm or similar information that should be kept confidential.
- Don't discuss holiday dates and schedules in detail or reveal when your house will be empty, where to find the spare key, or where you've stored all those Christmas gifts.

- Don't send images of nude or semi-clad people – particularly children – as they may end up on an offensive web site.
- Basically, don't send anything that you wouldn't want someone other than your recipient to read!

Things to remember

- Keep e-mail messages short. If you need to send a lot of information, attach it as a separate document.
- Don't send confidential or security-related information.
- Compress all images and attachments.
- Keep it simple – don't bother adding addresses, dates etc, to the top of your e-mail.
- Remember to add your name and any contact information.
- Try to answer e-mail within 48 hours.
- Be polite and avoid ambiguity, particularly in humour.

Chapter 7

Texting – Short and Sweet

Generally speaking, texting is between mobile or cellular phones. Texting is a great way to get a message across quickly, quietly and effectively, particularly when you are on the move, but it isn't suitable for every occasion. We look at when a text is best, and when it is definitely not, before looking at how to send them, what to say and how to make sense of all those abbreviations and smileys.

What is texting?

Texting is the sending and receiving of short text messages using the satellite technology of today's mobile phones. It allows you to communicate with other mobile phone users quickly, quietly and in writing, and is often cheaper, more discreet and more convenient than a voice call on the same handset. It is as portable as your mobile and has even helped save lives in mountain and marine accidents, as well as giving us the latest sports scores and allowing us to communicate in the noisiest of pubs and clubs.

When is texting useful?

Texting is ideal for:

- Getting facts and figures to someone in a meeting.
- Getting someone's attention when they can't speak on the phone.
- Discreet reminders.
- Last-minute changes of plan or location.
- Updates on ongoing situations.
- Cheap international communication.
- Getting a message through in a noisy place or where the telephone signal is weak.
- Chatting when you're far apart.
- Quick congratulations, greetings or apologies.

However, you should never text:

- When in a hospital, doctor's surgery, fuel station, airport or aeroplane.
- In sensitive situations – texting is even more frequently misinterpreted than e-mail.
- In examinations or the classroom.
- While driving or operating machinery.

Always remember that SMS means short message service – it is limited to 160 characters – so for lengthy messages or explanations use e-mail instead.

How do I text?

> The text message options for mobile phones vary tremendously, so you'll need to read your individual instruction book for detailed instructions, or check out www.text.it/howtext/index.asp, which gives instructions for all major phone/provider combinations.

In order to text you first need to know, or have stored in your mobile, the phone number of the person to whom you want to send a message. Then:

1 Use the menus on your mobile to find the 'Messages' option and select the option to write a message.
2 Use the buttons on the keypad to key in the words you want to send. Most buttons will have two or more characters associated with them – for example, 1 is often used to key punctuation, 2 to key a, b, c and all the various accented versions of these three letters. So you would press the 2 key once to produce an a, twice to produce b, and so on. If you leave more than a one-second gap between button presses, the mobile assumes you have moved on to the next character in your message.

   The 0 key is used to insert spaces between words and there are usually menu options available to insert smileys, symbols, numbers and so on. You can switch capital letters on and off as required or use lower-case only – this is often controlled by the # key.

   Practised texters often abbreviate everything to save space (and keying!) but this will come with time – some tips are given on page 98.

Predictive texting

Predictive texting is a system by which the phone appears to 'learn' what you want to type. It can be very handy if you use the same words frequently. But otherwise it can be a real nuisance, suggesting that you want to write a particular word when in fact you mean something completely different. If you find this happening when you first start to text, turn off predictive texting – see your manual for details.

3 When you are happy with the message, select the menu option to send and either type in the phone number or select it from your phone list. Select send again and the message will be sent directly to your recipient.

Is it instant?

People often make the mistake of assuming that texting is always instant. Often it is – but this is not necessarily the case. If your recipient has their phone switched off, or on silent, or is in an area of bad reception, it can take some time before they are able to read your message.

Some systems have an acknowledgement facility that tells you when your message has arrived, and when it has been read. Check your instruction manual for details.

Typical text messages

Here are some examples of text messages, both good and bad.

Message		Message	
Gerry, can you e-mail me the sales figs you quoted in last Friday's presentation please. Thanks very much.	✗	Pls e-mail me sales figs from Fri's mtg. Thx.	✓
done (in response to a message received)	✗	figs on way	✓
mtg canteen 12	✗	Pls meet me in the canteen at 12	✓

- The first message is just too long, requiring the reader to scroll through it, but the second says the same succinctly (note the use of understandable abbreviations) and politely.
- The third message could be confusing if you've sent or received a number of messages to one another: What has been done? The fourth message makes this clear.
- The fifth message shows how sometimes a quite simple request can come across as terse or even threatening. The sixth is far better.

Abrv8d msgs

The use of abbreviations in text messages is widespread and sometimes confusing. Unless you know your recipient well, and they know what you will mean, it is often better not to abbreviate too much.

But some abbreviations are easily understood and do save a lot of typing. For example:

2moro	tomorrow
2da	today
18	late
cu 18r	see you later

thx	thank you
pls	please
mtg	meeting
ph	phone

If you've experience of internet chat rooms or newsgroups, you are probably familiar with many of the other abbreviations used in text messaging. Otherwise you may well need a guide. Some useful books and web sites are listed in Chapter 12. Meanwhile, here are some typical abbreviated texts:

Message	Meaning
18 mtg cu ab7	I've got a late meeting – I'll see you around 7pm.
I lk u wan2 tlk?	I like you – do you want to talk?
hrd yr nws uok?	I heard your news – are you all right?
pls txt apr figs asap	Please text me the figures for April as soon as possible.
cu tonys @ 12	I'll see you at 12 o'clock at Tony's place.

Message	Meaning
ily	I love you!
SB d8in JW!	(Someone whose initials are) SB is dating JW.
cn u clct kds 2nt?	Can you collect the children this evening?
stk m25 wlb 18	I'm stuck on the M25 motorway and will be late home.
jan bdy 2409	Jan's birthday is the 24th September.
afaik mtg off	As far as I know the meeting has been cancelled.
cw82cu	I can't wait to see you!

Space dot dash

Unless your recipient is a particularly experienced texter, always use spaces between words and appropriate punctuation – it's not much extra effort and it really does make the difference between total gobbledegook and an understandable message. For example, which messages look clearer to you – those on the left or the right?

Plsrm2brsuit	pls rmbr 2 brg suit
Cuians8ish	c u @ ian's 8ish
mumbdywedgetprsnt	mum's bday on weds, pls get presnt.

Smile and the world smiles with you

'Smileys' or 'emoticons' are little pictures made up from one or more characters on your keyboard or mobile phone keypad. They resemble faces or facial expressions – sometimes whole figures or other objects – and they help to get over the problem of your reader not being able to see your expression or hear the tone of your voice. For example, a simple ☺ indicates you are happy, while a tongue-in-cheek comment can be easily identified by the symbol ;-^) (turn the book sideways if you can't see the face with its tongue firmly in cheek). Without this, the comment may be misinterpreted.

Here are a few of the most common smileys:

:-)	I'm happy
;-)	I'm winking (joking)
}:-*#	I'm very angry (swearing)
:'-(	I'm crying
:-0	I'm amazed (or singing!)
:-|	I'm straight-faced (or unimpressed)
{:%	I'm confused

There are many, many more, which you can find listed in the books and web sites given on page 185.

Some phones have an Add Symbol function that allows you to pick the most common smileys from a list to add to your message, rather than typing each one individually. Check your phone manual.

The future is text

We are only at the start of the mobile phone revolution. New advances in technology mean that pictures, photographs and documents can be whizzed from one handset to another in microseconds across the world and that the fiddly multiple key presses will soon be replaced with a mini keyboard. Texting is now more popular than any other form of communication for everyday use, and as it has a great youth following it is only to be expected that it will take over from e-mail and possibly even letters in the long term. So don't be afraid, get those fingers talking!

Tips for better texting

- Use smileys to indicate your mood.
- Don't use block capitals – it's the equivalent of SHOUTING!
- Use numerals for to/too (2), for (4) etc – e.g., 4ward, 2day, 2moro.
- Don't over-abbreviate or use ambiguous acronyms.
- Use abbreviations only when texting someone you know well – for that initial text spell out for extra clarity.
- Don't waffle, keep it short and to the point.
- Use spaces to avoid confusion and irritation.
- Don't bombard anyone with too many messages.
- Use the Reply function when answering a message you've received, so the recipient can refer back to the question.
- Be polite, especially when texting strangers. People have been sued for text abuse when making inappropriate or unwelcome remarks.
- When in a particularly quiet or noisy place, set your phone to vibrate rather than ring when a text message arrives.

Chapter 8

Junk That Junk Mail!

Does your mailbox overflow with messages, offers and information that you really don't want? You're not the only one. Unsolicited letters, spam on your computer, infuriating text messages coming from heaven knows where, even offensive or nuisance offers, we all get them in increasing numbers as vast computers holding databases of names and addresses just pump out all that paper and send it out. What do you do about it? How do you stop all that mail from cluttering up your life?

The organisations named below (whose addresses are given on pages 186–187) are primarily for UK readers. For information on how to stop junk mail elsewhere, contact your local library or citizen's advice centre, or run an e-mail search using terms such as 'junk mail', 'spam', 'unsolicited messages', etc.

Remember that offensive or threatening mail should be reported to the police immediately.

Stopping junk mail at source

To stop unsolicited mail from organisations you haven't previously dealt with, contact the Mailing Preference Service (see page 186 for contact details). They will store your details for up to five years, and organisations that sign up to their code of practice will be asked to remove your details from their mailing lists. Note that they cannot enforce this against companies with whom you have dealt in the past – or their subsidiaries (see below for help with this).

It will take about four months before you see a genuine reduction in the mailings you receive.

To stop unsolicited e-mail and text messages, write to the same address or sign up at www.e-mps.org.uk

If you are getting annoying text messages, remember that some mobile phone companies have software in place to intercept them. Contact your supplier and ask whether they can help. You can do the same with your internet service provider to help stop e-mail spam, but it is rarely totally effective.

Reporting nuisance or offensive mail

The following organisations deal with nuisance mail that is not covered by the preference services (for example, non-addressed flyers), and offensive advertising. This often originates from abroad, so can be difficult to regulate.

- For paper-based junk: If repeated or offensive, try the Advertising Standards Authority. They are responsible for prosecuting companies who ignore requests to stop sending mail or whose advertising is misleading or offensive. Alternatively, contact the Information Commissioner, who is responsible for upholding the Data Protection Act.
- For junk e-mail: It is best to tackle this through the internet, as many sites have been set up specifically to deal with this problem. Spamcop (www.spamcop.com) is one of the most respected anti-spam companies, but there are several others worth a try: run an

internet search on 'stop spam' to find those appropriate to you. Always report spam to your internet service provider and, where possible, to the provider used by the spammer. Don't bother replying directly to a spam message, as this often results in more, rather than less, spam.

- For junk texts: If you are still receiving nuisance texts after registering with the SMS Preference Service and contacting your mobile phone provider, send details to the Information Commissioner. Alternatively, if you can identify the company sending the message, contact their customer services department and ask for them to stop. Putting this request in writing will help with any further action.

Just ask!

Some of the most annoying mail comes from companies you have spoken to or purchased from in the past. A simple speculative request for a brochure can result in years of unwanted mail, from catalogues to flyers, some seemingly unrelated to your original request. To stop these, a simple letter is often enough. See the example on page 138.

If this does not work, you may have recourse to legal action, so make sure you keep copies of each letter you send and those you receive.

Getting your own back

If you are still receiving a lot of junk mail, try these tips:

- Credit card and loan companies sell your name the most frequently. Call them and ask them to stop. You will need your account details to hand.

- 'Free prize draw' and 'free gift' offers are usually just ways to get your name on a mailing list. Avoid these if you don't want the mail – the prizes are rarely worth having anyhow.
- Product guarantee or warranty forms often contain questions on your habits and income. Why? So they can target you for junk mail. In fact, these cards are rarely required to make a warranty claim, as long as you have your original receipt.
- When dealing with companies by phone, ask them to mark your account so that your name is not traded or sold to other companies.
- Whenever you send your personal details to any company or charity, write in large letters 'Please do not add my name to your mailing list.' This works for most reputable organisations.
- To avoid those advertising flyers and free newspapers that aren't distributed by the Post Office but by some teenager earning less than the minimum wage, just stick a card above your letterbox stating 'No free newspapers or advertising please.' It usually works, and if not you can stop them the next time and point it out.
- Return the reply envelopes (without postage) with a letter asking the company to desist.
- If you receive repeated mailings, save them over a period of weeks and return in one package (postage unpaid) with a letter asking the company to remove you from their mailing list.
- Return unopened junk mail in a large box, with a clearly printed letter asking them not to send any more. This works particularly well with small companies, as they often have to collect the parcel themselves from the post office – they won't want to do that too often!
- Publicity works wonders. Write, or threaten to write, to consumer groups/champions. Some addresses are given on pages 186–187.

Chapter 9

Rules to Remember

Use these checklists each time you sit down to write a letter, to ensure you communicate clearly and effectively, and get the desired response.

1 **Think before you write:** What are you aiming to achieve? How can you get the response you want and seem to be making a reasonable request?
2 **Use appropriate stationery:** Keep the flowers, cute kittens and snazzy borders for friends and family – and then only for newsy or routine letters. Use pre-printed stationery for invitations or create your own on a computer. Use plain, high-quality paper for everything else, in an appropriate size. For occasions, send a card with a short letter on a separate sheet. Use plain white or manila envelopes for all but personal messages.
3 **Keep it short:** If your letter is too long it is less likely to be read or acted upon immediately. Try to keep it to a single sheet of paper.
4 **Be polite:** Even if you're seething inside, don't insult or offend: this won't help you get what you want. Don't overdo the grovelling, but stick to simple, courteous language. There are very polite ways to get your contempt, dislike or disgust across!
5 **Get the facts right:** Don't embellish or exaggerate. Include details of all relevant dates, transactions, telephone calls, etc, and provide evidence where required (such as receipts, invoices, etc).

6 **When you want something done, give a deadline:** This helps both you and the person you are addressing to know what is required and when. It is also useful if any further action is required, for example, an attempt at redress through the courts.

7 **If it's personal, keep it personal:** Think about what your reader really wants to hear about – single men may not be too interested in the potty-training dilemmas of your youngest, but it'd raise a laugh with a fellow mum. Auntie Mabel will always coo over baby pictures but probably won't be that interested in the ins and outs of your high-tech job. Tailor the subjects to suit, and use language that your reader uses and understands rather than aiming for perfect grammar.

8 **Apologise – but only when necessary:** Don't say 'Sorry' unless you need to – never apologise for making a justified complaint – but apologise immediately and simply (don't gush) when you have made an error or caused offence.

9 **Be firm:** This particularly applies when you are complaining, asking for action or asking someone to desist in their action. Don't waver in your intent: don't, for example, apologise for not wanting to correspond with them or for making a complaint. You have the right to do so!

10 **Use the most appropriate medium for the job:** Don't text a complaint to your builder: it's too informal and easily lost; don't e-mail to a web site and expect an instant response; always use a formal letter for issues pertaining to your job. If in doubt and you have the time, always write a letter – it's rarely wrong.

Chapter 10

Writing to Titled People

When writing to royalty, government or local government officials, clergy and other titled persons, it is important to use the correct form of address and to start and end your correspondence correctly. If you need more information on the recipient, telephone the person's secretary or PA. For American equivalents, including local and state government and military titles, see page 184.

Royalty

When writing unsolicited letters to royalty it is normal to address your letter to a private secretary, equerry or lady-in-waiting. Address the letter to 'The Private Secretary to ...', and begin the letter 'Dear Sir'. Close with 'Yours faithfully'. However, if you want to address your letter directly to a member of the royal family, use the forms shown below.

Address the Queen as: Her Majesty the Queen
Begin: Madam, with my humble duty *or* May it please Your Majesty
End: I have the honour to remain (or to be), Madam, Your Majesty's most humble and obedient servant *or* Your Majesty's faithful subject
Address a prince/princess as: His/Her Royal Highness, the Prince/Princess of (title) *or* His/Her Royal Highness, Prince/Princess (forename) *or* (if a duke or duchess) His/Her Royal Highness, the Duke/Duchess of ...

Begin: Sir/Madam
End: I have the honour to remain (or to be), Sir/Madam, Your Royal Highness's most humble and obedient servant *or* Your Royal Highness's most dutiful subject
Address a duke/duchess as: His/Her Grace the Duke/Duchess of …
Begin: My Lord Duke/Dear Madam
End: I have the honour to be, Your Grace's most obedient servant *or* Respectfully

Knights and peers

Address peers other than a duke (Marquis, Earl, Viscount, Baron and Lord): The Most Hon the Marquis of … *or* The Rt Hon the Earl/Viscount/Lord of …
Begin: My Lord
End: I am, sir, your obedient servant *or* Yours faithfully
Address spouses of peers other than a duke: The Rt Hon Lady (surname) *or* The Rt Hon the Marchioness/Countess/Viscountess/ Lady of …
Begin: Dear Madam
End: I am, Madam, your obedient servant *or* Yours faithfully
Address a baronet: Sir (forename surname) Bt
Begin: Dear Sir
End: Your obedient servant *or* Yours faithfully
Address the wife of a baronet: Lady (surname only)
Begin: Dear Madam
End: Your obedient servant *or* Yours faithfully
Address a dame: Dame (forename surname)
Begin: Dear Madam (*or* Dear Dame (forename))
End: Yours faithfully

Address a knight of the realm: Sir (forename surname) (with appropriate letters after the name, e.g., KCB)
Begin: Dear Sir
End: Yours faithfully
Address the wife of a knight: Lady (surname only)
Begin: Dear Madam
End: Yours faithfully

Religious leaders

Address the Pope: His Holiness the Pope
Begin: Your Holiness (or Most Holy Father)
End: (if writer is Catholic) I have the honour to be Your Holiness's most humble/devoted/obedient child, (if writer is non-Catholic) I have the honour to be (or remain) Your Holiness's obedient servant
Address an archbishop: (Anglican) The Most Reverend the Archbishop of … (Anglican – Canterbury or York), The Most Reverend and Rt Hon The Lord Archbishop of …, (Catholic) His Grace the Lord Archbishop of …
Begin: My Lord Archbishop *or* Dear Archbishop
End: Yours faithfully
Address an archdeacon: The Venerable (forename surname), Archdeacon of …
Begin: Dear Archdeacon *or* Venerable Sir
End: Yours faithfully
Address a bishop: (Anglican) The Right Reverend the Lord Bishop of …, (Catholic) His Lordship the Bishop of …
Begin: My Lord Bishop *or* My Lord
End: Yours faithfully

Address a cardinal: His Eminence (forename) Cardinal (surname)
Begin: Your Eminence
End: Yours faithfully
Address a dean: The Very Reverend (forename surname), Dean of ...
Begin: Dear Dean (surname) *or* Very Reverend Sir
End: Yours faithfully
Address an imam or mullah: Because Islam has no formal hierarchy there is no specific guideline on how to address the leader of a mosque (the *imam*) or one of its teachers or scholars *(mullahs)*. However, the following is unlikely to give offence: Imam (forename surname) or Mullah (forename surname)
Begin: Most righteous Imam *or* Mullah (surname)
End: Yours faithfully
Address a rabbi or chief rabbi: Rabbi (forename surname) *or* The Very Reverend the Chief Rabbi (forename surname)
Begin: Dear Rabbi (surname) *or* Dear Chief Rabbi
End: Yours faithfully
Address a vicar, priest or rector: (Anglican and Catholic) The Reverend (forename and surname) (initials of any order, if appropriate)
Begin: (Anglican) Dear Sir/Madam *or* Dear Mr/Mrs (surname), (Catholic) Dear Reverend Father
End: Yours faithfully

Diplomats

Address a British ambassador: His/Her Excellency (rank) HBM's Ambassador and Plenipotentiary
Begin: Sir, My Lord etc, according to rank *or* Your Excellency
End: I have the honour to be, Sir, Your Excellency's obedient servant

Address a governor general or governor: His/Her Excellency (forename surname) Governor General of …. *or* His/Her Excellency (forename surname) Governor of ….
Begin: Sir *or* Your Excellency *or* My Lord (if a peer)
End: I have the honour to be, Sir (*or* My Lord), Your Excellency's obedient servant
Address a consul, vice consul or consul general: (Full name) Esq. HBM's Consul General
Begin: Sir
End: Sir, I am your obedient servant

Government and local government

Address the Prime Minister or minister: This is addressed by appointment only, for example: The Prime Minister, The Secretary of State for …
Begin: Dear Sir/Madam
End: Yours faithfully
If the writer knows the minister concerned, begin with the appointment (Dear Prime Minister) in which case end Yours sincerely
Address a member of Parliament: Mr/Mrs or Ms or other title (full name) MP
Begin: Dear Sir/Madam *or* Dear Mr/Mrs/Ms (surname)
End: Yours faithfully
Address a privy councillor: The Right Honourable (forename surname) MP
Begin: Dear Sir/Madam
End: Yours faithfully

Address a lord mayor: The Right Honourable the Lord Mayor of … (only to Lord Mayors of London, York, Belfast and Dublin) *or* The Right Worshipful the Lord Mayor of …
Begin: My Lord Mayor
End: Yours faithfully
Address an alderman: Alderman (followed by any title or rank and full name)
Begin: My Lord, Dear Sir, Dear Madam *or* Dear Alderman (according to personal rank)
End: Yours faithfully
Address a councillor: Councillor (followed by any title or rank and full name)
Begin: Dear Sir, Dear Madam *or* Dear Councillor (according to personal rank)
End: Yours faithfully

The judiciary

Address the Lord Chancellor/Lord Chief Justice: The Right Honourable The Lord Chancellor/Lord Chief Justice
Begin: My Lord
End: Yours faithfully
Address the Master of the Rolls: The Right Honourable Lord (surname)/Sir (surname), Master of the Rolls
Begin: Dear Lord/Sir
End: Yours faithfully
Address a High Court judge: The Hon Mr/Mrs Justice (surname) *or* Sir/Dame (surname). Note that for women judges, Mrs is always used even if the judge is unmarried

Begin: Dear Sir/Madam *or* Dear Sir/Dame (forename surname)
End: Yours faithfully
Address a circuit judge: His/Her Honour Judge (surname)
Begin: Dear Sir/Madam *or* Dear Judge
End: Yours faithfully

Academics and medics

Address a professor: Professor (forename surname)
Begin: Dear Sir/Madam or Dear Professor
End: Yours faithfully
Address a PhD: (full name) (initials identifying doctorate, e.g., DD, LLD)
Begin: Dear Sir/Madam *or* Dear Dr (surname)
End: Yours faithfully
Address a doctor of medicine: Doctor (forename surname) *or* (forename surname) MD
Begin: Dear Doctor *or* Dr (surname)
End: Yours faithfully
Address a surgeon or consultant: Mr/Mrs/Miss (forename surname), although this may vary outside England and Wales
Begin: Dear Mr/Mrs/Miss (surname)
End: Yours faithfully
Address a dentist: (forename surname) DDS *or* Dr (forename surname)
Begin: Dear Dr (surname)
End: Yours faithfully

Chapter 11

Sample letters

Here you will find sample letters (and a few e-mails) covering all sorts of subjects. Use them to get ideas for your own letters and to see how best to structure a letter on a particular topic, or just adapt them to suit your own needs. Use the paragraph spacing and general tone as a guide and decide on fonts to suit yourself.

Invitations and announcements

Informal social event

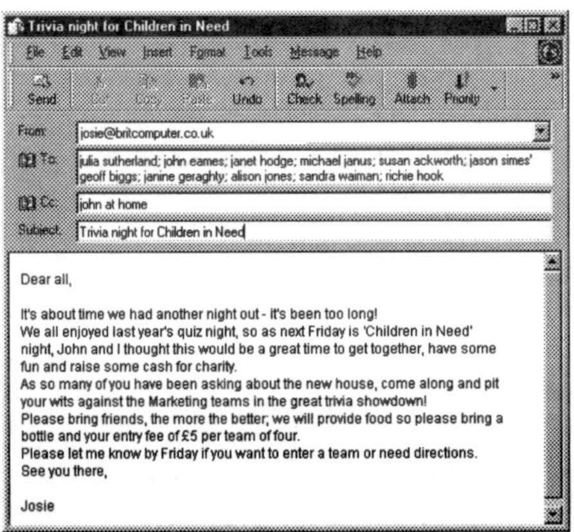

Trivia night for Children in Need

From: josie@britcomputer.co.uk

To: julia sutherland; john eames; janet hodge; michael janus; susan ackworth; jason simes' geoff biggs; janine geraghty; alison jones; sandra waiman; richie hook

Cc: john at home

Subject: Trivia night for Children in Need

Dear all,

It's about time we had another night out - it's been too long!
We all enjoyed last year's quiz night, so as next Friday is 'Children in Need' night, John and I thought this would be a great time to get together, have some fun and raise some cash for charity.
As so many of you have been asking about the new house, come along and pit your wits against the Marketing teams in the great trivia showdown!
Please bring friends, the more the better; we will provide food so please bring a bottle and your entry fee of £5 per team of four.
Please let me know by Friday if you want to enter a team or need directions.
See you there,

Josie

Engagement

Engagement

It is with the greatest pleasure that Brian and May Althop
announce the engagement of their daughter

Alice Althop

to

Harry Greenford

Join us all for a celebratory drink at
The Harvesters, Copnor Road on Friday 12 July from 8pm.

Jonathan and Rachel

invite you to a party to
celebrate their engagement

Come to
27 Waterside Gardens, Buckley
on Thursday 20 August
from 7:30 onwards

See you there (no RSVP necessary) Casual dress; bring a bottle!

Wedding

Note that there are specific ways in which to address invitations issued by divorced, separated, widowed or remarried parents; the couple themselves; or others who are hosting the wedding. Refer to the reference list for details of wedding web sites and publications that can help with this. This is the traditional, formal style.

Mr and Mrs John Sloford

request the pleasure of your company

at the marriage of their daughter

Andrea Ruth

to

Joaquin Lorenzo-Parier

at

St Mary and St Mark Parish Church, Hungerford

on Saturday 18 November at 3pm

and afterwards at

The Moat House Hotel, Marlborough

RSVP (by 10 October please) to:

Breezefields
Staunton Way
Hungerford
Berks
NB7 6SG

Birthday – child's

INVITATION

Alex Jones would like James to come to a party
to celebrate his fifth birthday.

It's at 18 Henpeck Close, Kingsford at 3pm on the 23 March.

Please let us know if you can come by calling 0182 612323
before 18 March.

See you there!

Jasmine is Seven!

Please come and help us celebrate at
Kids R Crazy, Belling Business Park, Moncham
on 27 July between 2 and 4:30

Note to Mums:
Please make sure your child is wearing socks and let us know of any special dietary requirements!

....

Jasmine,

☐ Yes I can come to your party ☐ Sorry, I won't be there

from ____________________

Return to 12 Joyful Way, Little Monch
or call us on 01821 8236125

Surprise party

This often takes the form of a letter, but a card format will do just as well – just make it plain it's a surprise.

Sssshhhh! Don't tell Sssshhhh! Don't tell Sssshhhh! Don't tell
!!! We're throwing a party for Jimbo !!!
It's going to be a surprise.
Please let us know if you're coming
and we'll provide the pies!
12th July at Funley Social Club; 7:30 if you want to surprise Jim
Dress to impress! RSVP to Jo on 01982 71122.

17 Blindman's Lane
COWLEY
Oxfordshire OX8 2US

12 September 2004

Dear friends,

November 2nd marks a very special day for our mum and dad, Jean and Jeremy – it's their golden wedding anniversary. Now Mum doesn't want a fuss, but we think that fifty years together must be worth celebrating!

They are off on a cruise for the big day to escape us, but we've other plans – a party at The Willow Tree restaurant in Cowley, with as many as possible of those who came to their wedding and lots of newer friends and family. The big day is October 20th, with a celebratory meal starting at 8pm. Guests are to arrive from 7pm to ensure a big surprise welcome for the happy couple.

Would you like to join us? If so, please let us know at the above address by 28 September and we will send you directions and a menu.

We look forward to seeing you – and remember, mum's the word!

Janice, Henry and Christine Allsop

Funeral

George Richard Duckton
1903-2004
May he rest in peace

You are invited to a service to commemorate George's life
at Hunford Parish Church
14 June 2004 at 2pm
and afterwards at
26 Honey Lane

No acknowledgement required.
Family flowers only: donations to The Healthy Heart Organisation
at George's request.

Emily Manning Ford
'Our darling daughter – taken from us so suddenly and so young'

Jean and Kevin Ford ask you to attend
the funeral of
Emily
at Stow Crematorium, 14 June 2004 at 2pm

Floral tributes welcome or donations to
Stow Hospital Children's Ward

Family round-ups

37 Quinter Road
FOLKESTONE
Kent
RH29 8UR

18 December 2004

Dear

I hope this letter finds you in good spirits and getting ready for Christmas. We Bellamys certainly are! It has been a very busy year for us all so we have decided to take a break and head off to Austria for some festive skiing.

It's been a year of many changes for us, so we certainly need the rest. John changed his job back in February and a good thing too – he's already been promoted and is really enjoying his new position. His office is closer to home too, so he's able to spend more time with us.

John junior and Stephen are looking good for the football team next year and have both done really well at school. They each now have a girlfriend to fill in all that extra time between matches and homework! Next year they'll be studying hard for exams, so are looking forward to making the most of the time they have left.

I've been blessed this year with a lot of time off, as I took my delayed maternity leave to spend more time with Rosie. She is becoming a very independent little toddler, walking and talking already and singing all day! I'd forgotten quite how joyful they can be at this age. We have great days shopping and she even helps me with the housework.

Well I'd better get back to the decorations. We all wish you a jolly and peaceful Christmas, and hope to see you next year sometime.

Best wishes,

Geraldine *John*
J xx *Steve*
and Rosie xx

Acknowledgements and thank-yous

Whenever possible, write these by hand. It shows you have taken the time and made the effort to thank people individually.

Wedding invitation (accept)

18 Honeyfield Gardens
Knighton
WESTBOROUGH
Hants
WB7 1US

18 January 2004

Dear Mrs Wulton

Thank you so much for your kind invitation to Jane's and Peter's wedding. I should very much like to come and look forward to meeting you on the big day.

Please would you forward me a gift list, or let me know where one can be viewed. Many thanks.

Yours sincerely,

Honor Grey

Honor Grey

Wedding invitation (decline)

18 Honeyfield Gardens
Knighton
WESTBOROUGH
Hants
WB7 1US

18 January 2004

Dear Mrs Wulton

Thank you so much for your kind invitation to Jane's and Peter's wedding. I should very much like to come but unfortunately I have already booked a holiday for that week so shall be unable to attend.

However, I would like to arrange a gift for the happy couple, so would you please let me know if they have a list, or if there is anything they would like?

Many thanks: I hope you all have a wonderful day and I look forward to seeing the photographs when I return.

Yours sincerely,

Honor Grey

Honor Grey

Thanks for a child

Some notes on children's thank-yous are given on pages 42–43.

The Woodlawns
Howton Way
GREATFORD
Devon
BD8 6ST

27 December 2003

Dear Bridget and Julian

Just a brief note to thank you so much for the children's wonderful Christmas presents. They were delighted with them all and have spent a long time playing with the drawing games especially. They've done some special pictures for you – I hope you like them!

We all had a great Christmas, traditional dinner and all. It made a lovely change from our usual skiing trip so we may do the same next year. Meanwhile we've the New Year to look forward to: my mother is having the children overnight so we can celebrate in style!

With best wishes for a peaceful 2004 – and thanks again,

Julia Horace

Julia and Horace

for Geraldine, Bradley and Joan

Acknowledgement of condolences received

Mrs Barbara Duckton
would like to thank you for your kind wishes on the recent sad loss of her beloved George.

It is comforting to know that he had such good friends.

On behalf of The Healthy Heart Organisation, many thanks for all your kind donations.

JOHN, SUSAN AND FAMILY

WISH TO THANK YOU FOR ATTENDING GARETH'S MEMORIAL SERVICE
AND FOR YOUR WARM AND COMFORTING WORDS ON THEIR LOSS.

WE HOPE YOU WILL ALLOW US FEW QUIET DAYS OF PRIVATE CONTEMPLATION
BUT LOOK FORWARD TO MEETING YOU AGAIN SOON.

Thanks for wedding gift/attendance

Flat 1B
263 Acacia Crescent
WALTON
Suffolk
IP6 8HW

29 February 2003

Dear Marcia and Jerry,

Thank you so much for the beautiful photo frames. We are looking forward to filling them with the 'official' wedding pictures as soon as they arrive. They will look perfect in our new flat, which is very modern in style.

It was wonderful to see you both at the wedding – it's been too long! – and we both hope you enjoyed the day. We had a fabulous time and a very relaxing week in the Seychelles, with great weather and lots of time on the beach.

Martin sends his best wishes. We'll have to meet up for a meal next time you're up from London: we've already found a great little place just up the road.

We look forward to seeing you both soon,

Angela Martin

Angela and Martin Howne

Letters to your community

Starting a neighbourhood watch group

17 Honeyford Way
SLOWTOWN
Beds
BD9 2US

18 March 2004

Dear friends,

It's a sad fact that local crime is on the increase, and particularly burglaries and car theft. Although our Slowtown Police do a wonderful job – they have one of the best clean-up rates in the county – they really could do with our help.

As secretary of the Cavendish Residents' Association, I'm writing to ask whether you'd join us in setting up a Neighbourhood Watch network for our estate to help beat crime. It will take up only a few minutes of your time each month, and requires no qualifications – just a sense of community and a keen eye and ear for suspicious activities in your local area.

Quite apart from the obvious benefit in preventing and reducing crime, and making it known that the estate is a no-go area for car thieves, joining the scheme will earn you a discount of up to 10% on your home and car insurance, so what have you got to lose?

I've enclosed a leaflet about the scheme which tells you how it operates and how you can help. If you need any more information, or if you'd like to be one of our area co-ordinators, you can call me or our designated PC, John Beeton, at Slowtown Police Station. I will be arranging a meeting early next month – details soon – and look forward to seeing you there.

Regards,

Val

Valerie Staunton
01826 1927361

Invitation to a coffee morning

You can style invitations as a letter or more like a party invitation (see page 116).

12 Brook Avenue
CHISLEFORD
Bucks
AY9 2JD

9 April 2004

Dear neighbours and friends,

We are holding a coffee morning in aid of the Chisleford Brownies on Saturday 23 April at 12 Brook Avenue, and would be very pleased to see you there any time from 10.00.

Coffee, tea and cakes will be available, and there will be games and fun for the children. We will be holding a raffle and children's lucky dip to boost the Brownies' Jamboree funds, and would warmly welcome any donations of prizes.

An entry fee of £1 (adults) or 50p (children under 16 – under fives go free!) will include both coffee/tea and cake, or squash and a biscuit for the children.

See you there!

Miriam Jules

Miriam and Jules Cooke

Asking neighbours to sign a petition

27 Bewley Way
CHARLESFORD
West Midlands
B12 9SK

April 2004

Dear neighbours and friends

As you may have seen in the *Evening Herald*, Charlesford Borough Council has closed Allegra Park following vandalism that resulted in some extensive damage to the toilet facilities. The Parks Department has told the *Evening Herald* (see attached) that they do not foresee the park being re-opened in the near future, as the cost of maintaining the park is just too great.

As a local resident and mother of three, I find this intolerable. Had the council provided the CCTV it promised at the end of last year and which was included in their annual budget (see attached) the vandalism and subsequent closure would not have occurred. Now our children have no safe communal area in which to play and no similar play equipment within walking distance.

Together with concerned parents from surrounding estates, I am organising a petition to be presented to the council, asking them to honour their commitment to install CCTV monitoring and reopen the park. Would you sign?

Copies of the petition can be found at all the outlets listed below, or you can sign and return the form attached to this letter. Many thanks for your support.

Yours sincerely,

Pearl Friston

Pearl Friston

Petitions can be signed at: Julie's Beauty Parlour, Aroma-therapy Florists, Handford News, Spar (Hilton Road), St Mark's Playgroup or Tiddlywinks Pre-school.

Organising a clean-up

18 Jonquil Way
TWAINWORLEY
Norfolk
NW9 0OS

18 January 2004

Dear friends,

What are you doing next Saturday afternoon? Watching the match? Sinking a few beers? How about doing something different and having a bit of fun?

The recent storms have washed and blown all sorts of rubbish into the canal, and it is time for a clean-up. We need volunteers to help us restore one of our local favourite walking spots to its former glory. Coffee and sandwiches will be provided free in return! All you need is a good pair of wellies or walking shoes, a waterproof jacket and old clothing; rubbish bags and tools will be provided.

The clean-up starts at 2.30 and we hope to be finished by 5.00 – the more of us that turn up the less time it will take! We will be meeting at Stirling Bridge where you will be assigned an area and a team to work with.

We ask that children under the age of nine do not join us, as they will need constant watching on the canal bank. However, older children would be very welcome – they often enjoy the atmosphere and the fact that they are helping their community. The 12th Twainworley Scout Group will be assisting with the refreshments, so many of their friends may already be there.

We look forward to seeing you there!

Many thanks

John Mansfield

John Mansfield
Canal Clean-up Group

Letters to your neighbours

Noise or annoyance issues

The following approach might work best with an elderly neighbour:

Juniper Cottage
Harland Wood
FEBRING
FY72 9SH

26 September 2003

Dear Mr Groening,

We haven't seen you out and about recently, so I hope this letter finds you well.

With the recent beautiful weather, we've all had a lot of our windows open in the evenings. Brian and I have noticed that your radio has been left on at high volume until quite late at night while the kitchen window is open. This being quite close to our fence, we are finding it rather intrusive and it is hampering my efforts to get the baby to sleep.

I realise your radio is a lifeline to you, but would you consider either reducing the volume or closing the window after 6pm? This would really help us enormously.

Thank you so much. Don't forget we are always here if you need anything.

Yours,

Julie Smith

Boundary issues

12 Benfield Gardens, etc.

12 June 2004

Dear George

As you've probably noticed, the hedge between our gardens is becoming woody and unattractive, having rather outgrown itself. As neither of us has the time to trim it regularly, Paula and I are wondering whether you'd agree to our replacing it with an easy-maintenance wooden fence.

The problem is, our title deeds don't show which property actually owns this boundary. Do you know? It doesn't matter to us, really, as we are happy to do the work if you agree, but we thought we'd best check first in case you'd any objection.

Assuming you agree, we'd need to get the work done before the rainy season. I have the first two weeks of September available, which would be ideal.

Would you like to call over for coffee over the weekend, so we can discuss it in more detail and look at the different styles?

With kind regards,

John

John

Asking a neighbour for professional advice

While you can request general advice from a qualified professional neighbour, you cannot expect him or her to provide for free information or work that would normally be paid for. This can lead to awkwardness, even if you hadn't intended it to be so. So make this clear in your letter.

12 Benfield Gardens, etc.

12 June 2004

Dear George,

I'm having a little difficulty with the local planning department over a proposed rear extension to 8 Shelfield Way, which will overlook our garden. With your experience in such matters, could you give me any advice?

At this stage we need only make a preliminary objection by the end of the month, but I need some general tips on how to approach this and who best to canvass for support.

If the objection fails and permission is granted I would like you to take on this matter for us on a professional basis. Hopefully it will not come to this, however.

I'll be at home any evening this week or if you'd prefer I can visit you at your office.

With kind regards,

John

John

Getting a loaned item returned

36 Frosty Way

2 December 2003

Dear Geoffrey

I can't find my snow shovel anywhere and I remember I lent it to you last year when yours was broken. Would you look it out before Friday so I can collect it, or I'll have to buy another as we're off to the cabin next week and the weather is turning.

Hope you, Sarah and the kids are well – we must all get together for lunch again before Christmas.

Thanks,

George

George

Condolences and letters of support

On bereavement

Manor Lodge
SOWNTON
Bucks
BH7 3YF

12 January 2003

Dearest Connie,

Donald and I were so saddened when we heard the news about Desmond. While we all knew the end was inevitable, I guess we hadn't realised quite how ill he had become in those final weeks. I am so sorry we didn't get to say goodbye in person.

We both had a great deal of affection for Desmond, especially for the way he took care of you after Johnnie left. He was always such a caring, thoughtful man. I'll never forget the time that he drove all the way to Campden to meet me after the train broke down, just so I wouldn't have to get on the bus. And he insisted on driving me all the way home again. He was one of nature's gentlemen and we will not forget him.

It must be a very difficult time for you but I'm sure your family are taking good care of you and sadly we must all find our own way through those dark days. Please do remember we are always here and when you feel like a break you are very welcome to come and stay for as long as you like.

With all best wishes, and our unbounded sympathy

Ethel Don

Ethel and Don

Let's say that someone you worked with, but didn't particularly get on with, has died in questionable circumstances. You need to acknowledge his family's grief with a certain degree of honesty, plus a good handful of tact. Don't pretend to have been the guy's best pal or try to comment on the manner of his passing: instead find something noteworthy in his character and draw on that. For example:

> 14 Jensu Road
> Havering
> Sussex
>
> 12 June 2004
>
> Dear Mr and Mrs Jones
>
> This must be a very difficult time for you and all those who were close to Ben.
>
> I worked with Ben for several years and always found him to be a man who enjoyed life and who was widely admired for his sharp wit and ability to close a deal. While we rarely met socially I know he had a wide circle of friends as well as a close family, and I am sure he will be much missed.
>
> Yours sincerely
>
> *David Newbourn*
>
> David Newbourn

A similar approach may sometimes be needed for letters of reference.

On relationship breakdown

12 Stormy Way
Benfield
ESSEX
SS82 7TS

12 September 2004

Dear Joe

Jane and I were sorry to hear from Faye that you two are splitting up. I know things have been difficult for you recently but I guess we all hoped they would resolve themselves. It's always a great shame when two people who you care about end up unable to live with one another.

As you know, Jane's been friends with Faye for a long time, but she also thinks very highly of you, as do I, and we certainly wouldn't like this to stop us from staying friends. We understand if you want to stay out of Faye's way for a while, but any time you need a friendly face or someone to talk to, just call me and I'll bring over some beers. If there's anything we can do to help with the practicalities of the house move, let us know.

Look after yourself and keep smiling.

Kev

Kev

Apologies

12 Brook Court
12 June 2004

Dear George,

I do hope I haven't upset you or Miranda: Paula tells me I made some awful joke on Saturday evening and that you didn't take it at all well. I'm very sorry: you know I would never set out to offend anyone, particularly such good friends. Will you forgive and forget?

With best wishes,

Jack

'Sorry' letters that give bad news can be particularly difficult to get right. Here it is better to give some background or reasoning behind the bad news before you actually get to it. For example:

52 Henley Road
4 December 2003

Dear George,

Thanks for your note about the snow shovel. I'm very sorry I hadn't returned it before now – I really had clean forgotten it.

I have spent the last two days clearing the garage looking for the skiing gear but I cannot find the shovel anywhere. I clearly remember putting it in there after the New Year snowfall, but it is nowhere to be found.

I'm really sorry but it looks like I'll have to get you a new one. I'm going out to the garden store on Friday – come with me and we'll pick out the one you want. Let me know what time would suit you best.

Thanks – and apologies.

Geoffrey

An e-mail may be appropriate for a business communication.

Saying no

If asked to take on a role or fill a position that you'd rather not, base your letter on the sample shown on page 179 that turns down the offer of a job.

Refusing to donate funds

Wildside
Howler Grove
BEDFORD
BD7 8ST

18 November 2003

Dear Anne,

Thank you for your letter of 26 August. I am sorry to hear of the plight of the fur seals; man's brutal inhumanity to such animals seems exceeded only by that toward his own kind.

I'm afraid I am unable to assist you with a contribution on this occasion, as I have already donated heavily this year to Oxfam and to aid the refugees in Zaire. However, I wish you all the best with your campaign and do let me know if I can help you in the future.

Kindest regards,

Donald Brump

Donald Brump

Refusing a request for help

If you do have to refuse a request, you may be able to suggest someone else the person could contact.

Flat 1B
Wolsey Court
EGGWORTH
Sussex
RN8 6ST

14 July 2004

Dear Melanie,

Thanks for your long letter – it's great to hear all your news, although obviously I am sorry to hear about Daniel. I hope the break heals cleanly and he's soon back on his feet.

I wish I could help you out with looking after the children as you requested, but to be honest my time right now is very limited. I have just started a new business and these are critical days – I am working day and night just to keep afloat. I'm sorry if this seems rather self-centred but I have put a lot into this and cannot afford to see it fail at this stage. However, if you need someone to pick up the shopping or just to talk to then call me. I will endeavour to visit Daniel over the weekend.

Meanwhile I've asked Mary to give you a call; she's not working at the moment and loves kids, and she says she'd be happy to act as childminder if you could collect her and run her home afterwards.

With love to all the family, and especially Dan.

Katrina
Katrina

Letters about unsolicited mail

Asking for junk mail to stop

24 Howley Way
GUNFORD
West Sussex
GH3 8WG
24 July 2003

Customer Services Department
Garden Gate Company
BIRMINGHAM
B28 9GV

Dear Sir or Madam,

Re: Catalogue Mailings

Some two years ago I telephoned your company to request a brochure showing your selection of garden railings. At the time I asked for my details not to be included on your mailing list.

Despite this request I am still receiving regular mailings from your company and its associates on all manner of garden and ornamental products. I have recently returned several of these, again asking for my name to be removed from your list. However, the catalogues continue to arrive.

I realise that you adopt direct marketing as a successful method of selling your product. However, I no longer have the need for these products and do not wish to receive any further information from you or the companies with whom you exchange information. Would you please, therefore, ensure that I do not receive any further mailings at this address.

Yours faithfully

Gordon D Brown

Mr G D Brown

Asking someone to stop writing

25 Stoney Way
HERTFORD
EN2 9SH

26 November 2002

Fraulein Johanna Konig
35 Jonnstitstrasse
NEUMARKEN 102846
Germany

Dear Ms Konig

I have today received another letter from you, despite my earlier requests for you to stop this unwelcome intrusion into my private life.

As you know, I am now a married man with a loving family, and I do not wish to resume our former 'relationship' in any way. It has been some 15 years since we met and I cannot truthfully say I have regretted that separation since. I do not reciprocate the feelings that you express in this or your previous letters, and I ask you again to drop this correspondence.

I have no grievance against you but merely wish to continue my life without complication. I politely suggest that you do the same. As I have now asked you to stop writing on three occasions, I must warn that any further attempt by you to contact me or my family will cause me to consult my solicitor regarding legal action on the matter.

Regards

J. Winton

John Winton

Asking to be removed from an e-mail list

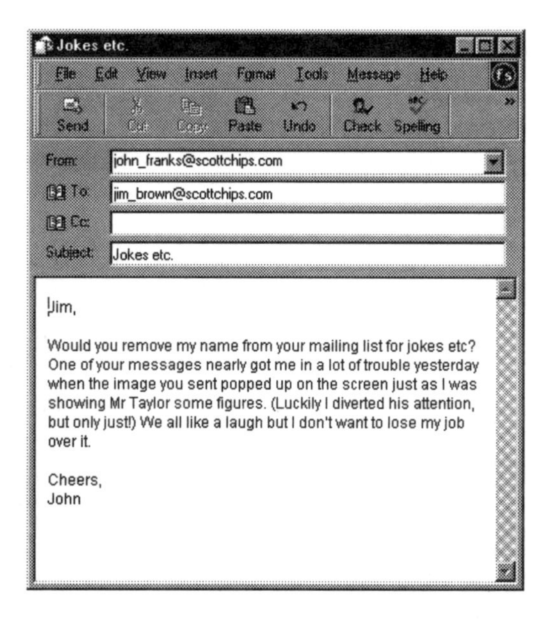

Jokes etc.

File Edit View Insert Format Tools Message Help

Send Cut Copy Paste Undo Check Spelling

From: john_franks@scottchips.com

To: jim_brown@scottchips.com

Cc:

Subject: Jokes etc.

Jim,

Would you remove my name from your mailing list for jokes etc? One of your messages nearly got me in a lot of trouble yesterday when the image you sent popped up on the screen just as I was showing Mr Taylor some figures. (Luckily I diverted his attention, but only just!) We all like a laugh but I don't want to lose my job over it.

Cheers,
John

Appeals for donations or bequests

18 Cudgen Grove
STYFORD
Herts
EN26 8ST

28 November 2004

Dear neighbours and friends,

The festive season is approaching and I'm writing to ask you to join me in ensuring that a few more of our townspeople can look forward to a great new year.

As you know, the Styford Village Hall was recently closed because its antiquated plumbing has become unusable.

One group that has not been able to find new accommodation is the Styford Stingers, our proud and accomplished wheelchair basketball team. What makes this doubly sad is that the Stingers were flying high in the Compton League this season and had their best ever chance of representing the country at the next Paralympics in 2006. As a result of the closure, they are now unable to qualify as they can only play in away matches.

The repairs to the Hall will cost no more than £2500, thanks to a generous offer by Churchill Builders to complete the work at cost. However, despite their best efforts, the players, their families and other interested groups have so far been unable to raise the funds required. We still need around £1400.

A renovated Hall would be available to all, so could I ask you to give a little extra to your community this Christmas and send us a donation to help us reach our target? Cheques can be sent to the address above, made payable to Styford Hall Appeal.

Thank you for your time, and a Happy Christmas to you all.

John

John Flanders

Houghton Lodge
CHURCHTOWN
North Yorks
YR27 8ST

26 April 2005

John Gunner Esq
Rupley Farm Estate
HONEFORD
North Yorks
YR28 0WT

Dear John

As chair of the Retired Farmers' Association, I am writing to ask for your help and compassion for your fellow farmers in these times of need. I know you take a great interest in our work and have been most generous in your help in previous years, and I am sure you realise the problems that we are all experiencing have a most drastic effect on those farms already struggling after the foot and mouth crisis.

During the crisis, volunteers and professional counsellors helped us maintain a vital lifeline to affected farmers with our 24-hour telephone support network. Generous donations from the public and members alike allowed us to maintain this support until recently, but we are now unable to pay the cost of the toll-free access or our helpers' expenses. Desperate and depressed farmers are no longer able to get that advice and emotional support that they need from those who understand.

Could I therefore ask you to help us with a donation – and some of your time if you are able? I've enclosed a leaflet about the service and the many ways in which you can help, and I look forward to hearing from you.

Remember, the service is provided by farmers for farmers, because only we know what farming today is like.

Yours sincerely

Henry
Henry Futton
Chair, Retired Farmers' Assn

Letters to the council

Requesting information on planning issues

14 Underwood Gardens
JOBTOWN
Herts
EN7 6YS
17 September 2003

Planning Department
Jobtown Borough Council
Civic Offices
JOBTOWN
Herts
EN8 7YW

Dear Sir or Madam

Re: Planning requirements for garage at 14 Underwood Gardens

I am hoping to build a garage on the side of the above property. For several years there has been a makeshift car port in the position in which I now wish to build a brick garage. The new building, basically the same size, will adjoin the house on its left side and the opposite wall is sited some 2m from the boundary fence. The proposed dimensions and design are shown on the enclosed drawings.

Do I need to formally apply for planning permission to build this?
What fees will I need to pay for such an application?
How soon could your planning inspectors make a decision on the matter?
Do I have the right of appeal if the application is rejected?

I would be grateful to receive your comments as soon as possible, as I hope to have the garage completed before the winter, if permission is granted.

Many thanks,

Henry Ford
Mr H P Ford
Enc. drawings/specs for garage

Objecting to planning application

6 Giorgio Avenue
HONITON
Devon
EX8 6TS

Your ref: HSIW/7SY/HB
18 June 2004
Planning Department
South Devon Borough Council
EXETER EX8 6YZ

Dear Sir or Madam

Re: Application to build commercial unit at 8 Giorgio Avenue

I would like to register my objection to the above plan, as per your letter of 15 June.

I have lived in this neighbourhood for some 12 years and it has always been a quiet, friendly area with mainly elderly residents. However, the new neighbours at No. 8 now propose to build a car maintenance business in their back yard. This will lead to: an increase in traffic and parked vehicles, leading to an increased risk of road traffic accidents; a vast increase in noise pollution during working hours, which I understand include Saturday; an increase in persons loitering in the immediate vicinity and the potential for car crime; inevitable pollution of the ground and water supply through leaking oils and lubricants and a decrease in the value of neighbouring properties, including my own.

From the plans I have viewed at the council offices, it also seems that the new building will be within one metre of the boundary with my property. This I view as totally unacceptable. There are plenty of industrial areas in Honiton that would be more suitable and I urge the committee to reject this application forthwith.

Yours faithfully,

J. Truman

Mrs J Truman

Letters to your MP or other representative

See pages 109–110 for details of how to address MPs and other local or national government representatives.

Asking your MP to attend/open an event

28 Marten Avenue
SHONTON
Derbyshire
DB8 2OD

12 March 2004

Madeleine Glover MP
Conservative Party Office
Handsville Road
DERBY
DB24 9SM

Dear Ms Glover

As president of the East Derbyshire Mothers' Union, I am writing to ask whether you would consider speaking to our members at our annual conference in May.

The theme this year is 'Getting back round the table', a reference to the changes in family life resulting in, or caused by, the demise of the communal family meal. As a working mother and former food writer our local members thought you ideal.

The conference is being held at Hamilton House, near Burnfield, from 12 to 15 May. An opening or closing speech would be best, if you are available.

Please let me know by 15 April whether you will be attending. If you would like more details, please telephone me on the number shown above.

Thank you.

Yours sincerely,

J A Trotter

Judith A Trotter (Mrs)

Asking your MP to raise an issue in Parliament

27 Slumber Crescent
Whitford
FAREHAM
Hants
PO71 6ST

16 April 2003

Charles Godwin, MP
Woodleigh House
Great Findean
FAREHAM
Hants
PO65 9YS

Dear Sir,

Re: Pollution of the River Whit by the Ministry of Defence

Further to recent reports in the local press, I am writing to urge you to act on the matter of pollution caused by the Ministry of Defence in this area, and to raise the issue with the Ministry and, if necessary, in Parliament.

As a keen angler I have noticed over the past nine months that the quality of the water in the river has steadily decreased, with frequent small oil slicks and surface foam and increasing amounts of dye being dispersed into the water. I initially believed this to be caused by factory effluent and notified the Environment Agency, but they found no leaks from local manufacturers and suggested the problem had arisen as a result of chemical tests taking place at the Shorten Down defence research facility, some four miles upstream.

I wrote to the authorities at Shorten Down in March, but have received neither acknowledgement nor answer to my queries. It seems I am not alone; several of my fellow anglers have done the same and all our letters have been ignored. The local press has taken up the story on our behalf but the Ministry of Defence has refused to comment on the situation.

This pollution is causing considerable environmental damage to the river and its embankments. Dead, poisoned fish are now being eaten by birds who are in turn dying and being eaten by other wildlife. Riverbank plants and water weeds are being suffocated in oil and the water is becoming stagnant. Soon the river will be past saving and another natural habitat will have been destroyed.

This problem must be addressed NOW. Please use your influence in the House of Commons and with the Defence Minister to ensure it is known about at the highest levels and that action is taken on behalf of your local community and its constituents.

Thanking you in anticipation,

Yours faithfully

Tom Shaw

Mr T N Shaw

Letters to the tax authorities

Querying the application of a penalty fee

28 Hinton Grove
CHALERFORD
Dorset
BH75 8GT

Your ref: 82736 82673K
4 February 2003

Inland Revenue Dorchester 'C'
Upley House
GROTFORD
Dorset
BH87 5RF

For the attention of Mrs Trotford

Dear Madam,

Re: Submission of self-assessment return

I have today received your letter of 2 February regarding the non-submission of my self-assessment tax return. I am concerned about this matter as I delivered this return, plus a cheque to cover the required income tax, to your office by hand on 26 January. I have attached a copy of the receipted slip given to me by your receptionist.

I have contacted my bank who inform me that the cheque (no. 0027362 for £1872.15) has not been presented to date. I therefore assume that the paperwork has been mislaid within your office.

Taking the above into account, I hope you will remove the fixed penalty of £100 that has been applied to my tax account for late submission, along with any interest that has accrued to date.

If you have any further queries, please telephone me on 01826 1927356.

Yours sincerely,

D Hole

Doreen Hole

Asking for advice on tax relief

18 Gormley Way
HOWARD
Nacshire
UB6 9SG

Your ref: 102917 291872K
16 March 2004

Inland Revenue
Customer Service Centre
Hunton Cross
GRIMSBY
GR9 8SJ

Dear Sir or Madam

Re: Tax relief for working parents

In the recent budget statement I noticed a mention of a working parents' tax credit. I was not previously aware that such a credit existed, and I would like to know whether I may be eligible to receive it.

My husband and I both work part time (20 hours per week each) in order to care for our twin daughters, aged four. Our combined income is around £16,500 per year including bonuses and tips. We are in receipt of a small amount of housing benefit but no other income support.

If you believe we may be eligible to receive this tax credit, would you please send us the application forms, plus details of how we could reclaim any relief due for the previous four years, if applicable.

If you need further details of our income, please contact me on 01827 827261 at any time.

With many thanks.

Yours faithfully,

Julia Robins

Julia Robins

Querying a tax calculation made on your behalf

19 Cavendish Place
Handley
BEDFORD
SH8 9UY

Your ref: 10298 37162K

12 May 2004

Inland Revenue
Tax Code Enquiry Office
Chadfield
BEDFORD
SH9 8TR

Dear Sir or Madam

Re: Tax code for tax year 2004/03

I have today received from you a notification of my tax code for the forthcoming tax year. A copy is enclosed for your reference.

I disagree with this coding for the following reasons:

I am currently a full-time student at Bedford University and thus unable to earn an income. I did advise your colleagues in the self-assessment office of this in September of last year (copy attached) but this seems not to have been taken into account in calculation of my tax code.

The tax code details include a figure for a benefit in kind of 'use of mobile phone'. I do not possess such a device, and never have done.

Would you therefore please review your assessment of my tax code and send me a revised figure as soon as convenient.

Yours faithfully,

Milly Montana

Mrs M. Montana

Writing to advise of a change of circumstances

You could also use this letter to advise banks, etc., of such changes.

4 Oak Road
ALSOP
Yorks
NO7 3SD

Your ref: 82736 92736 K

29 September 2000

Inland Revenue
Customer Records Department
Hadleigh Tower
IPSHOT
Kent
IG9 8JA

To whom it may concern

I have recently divorced and have returned to using my maiden surname since moving to my new address. Would you please amend your records to reflect this. I enclose a copy of the decree absolute indicating my change in marital status and a copy of my new tenancy agreement showing my name and new address.

Please ensure my records are dissociated from those of my ex-husband with immediate effect and that all correspondence concerning my own tax records is sent to my new address.

I would appreciate it if you would confirm receipt of this letter in writing.

Yours faithfully

K J Hutsford

Karen J Hutsford (formerly Knott)

Letters to schools and the education authorities

Requesting a school place

41 Crown Crescent
FOLKESTONE
Kent
MG7 6ST

12 May 2004

Education Department
Kent County Council
Cathedral Close
CANTERBURY
CN7 5ST

Dear Sir or Madam

Re: Seeking a school place at High Lane Juniors

We have recently moved to the above address from outside the county and wish to apply for a place at our local Junior school for my son Callum, aged 7 (d.o.b. 13 July 1995). I am told by the school that the year group is currently full but that applications can be made to your office for next term. I enclose the forms given to me by the school, duly completed.

I realise that numbers are restricted in all local schools but I urge you to allow my son to attend this school. I am a widowed parent with two other pre-school children, and I have no transport with which to take my son to a school further away. To send him elsewhere would seem detrimental to his social needs and would cause great difficulty for us as a family. If you are unable to accommodate Callum at this school immediately, I hope it will be possible to place him on a waiting list and for you to provide transport to another suitable school.

Yours sincerely,

Mrs Olivia Green

Enc. Application forms ED/A6 and ED/B5

Requesting a meeting with a teacher

18 Playford Way
THURSTON
Lincolnshire
LC8 6ST

22 May 2004

Mrs Angela Chinley
Thurston Infants School
School Lane
THURSTON
Lincolnshire LC8 2UI

Dear Mrs Chinley

Re: Ellie Brown

I would like to meet with you in the next few days to discuss some problems that Ellie appears to be having at school. Would you let me know a date and time that best suits you?

Over the last four weeks, Ellie has been unwilling to go to school in the mornings, which is very out of character and is worrying me a great deal. She has become quiet and withdrawn and something appears to be troubling her. She is unwilling to discuss it with me or her sister but it appears to be connected to her classmates, so I am hoping you may be able to shed some light on this for me.

In particular, I am wondering whether it is connected with the disappearance of Ellie's lunch money, which we discussed at our last meeting in April. I have again received a notice from the catering staff that the money has not been received for the past three weeks, but it was definitely sent with Ellie each Monday morning.

I look forward to hearing from you, and thanks for your time.

Yours sincerely,

P Brown
Mrs P Brown

Asking for advice on your child's education

18 Fortune Green
IPTON
Cheshire
MN6 7ST
27 August 2004

Mr Bob Frame
Mathematics Department
The Ipton Grammar School
Chancery Grove
IPTON
Cheshire
MN6 8SI

Dear Mr Frame,

Re: Edward Freiling – Trinity Two

Further to our lengthy discussions on the matter last term, we would like to seek further advice from you on Eddie's abilities in mathematics, particularly now he is entering a new year group.

In particular, we remain concerned at Eddie's slowness at complex numbers and algebra. He is still struggling in this area, despite the extra maths tutoring that you recommended, which he receives twice a week, in addition to your own efforts and ours. Eddie is becoming frustrated at his inability to comprehend or calculate the sums set and this is affecting his confidence in other areas.

We would like to know whether you have any further suggestions on how we can help Eddie with his maths, and how the problems he is having will affect his progress through the school. I know maths is a core subject at Ipton Grammar, and is considered of prime importance for boys wishing to go on to a technical career.

Do you now believe Eddie to be 'past help' in this area? Is it worth our investing in further, costly tutoring or is this just a 'blind spot' for him? If so, is this likely to impact on other subjects? Eddie wishes to become a civil engineer, but we are concerned that the problems he has with maths will prevent this; if so we'd rather know as soon as possible so as to steer him in another direction.

Paul and I very much appreciate the help you are giving Eddie and us in this matter. You are a fine example of a helpful, concerned and patient teacher, and we have expressed this with high recommendations to your headmaster.

Yours sincerely,

Jan Freiling

Mr & Mrs P Freiling

Letters to the media

Making a complaint about coverage or content

27 Hunstable Way
JOHNSTOWN
Leith
SU2 9SN
12 December 2004

Maggie Chissock
'Viewers Bite Back'
HighScot Television Corporation
DUNFILMIN
Galloway
DF9 7ST

Dear Maggie

I am writing to express my shock at the recent broadcasting of 'The Man with Five Wives' on HS1.

This was a semi-pornographic film, with repeated sexual language, yet was shown on a Sunday afternoon in late November, a time when many families are sitting around the television to avoid the Highland weather. What can your schedulers have been thinking of?

I know this was a late replacement for the cricket, but surely more care should be taken for weekend emergency programming. I faced some very awkward questions from my pre-teenage children, who channel-surfed on to it while I was out of the room.

I hope the corporation will broadcast a public apology and immediately review their procedures for such late additions.

Yours sincerely

Muriel Blagg

Mrs M Blagg

Asking for coverage of an event

Jasmine Cottage
FERNWALL STONEY
Nottinghamshire
NT21 9SY

2 September 2003

Daily Echo
PO Box 201
NOTTINGHAM
NT1 1EC

Attn: Sports Desk

Dear Sir or Madam

I am writing to ask for your assistance in publicising the Three Fernwalls Charity Golf Cup, being held at the Fernwall Norton Golf Club on 2–3 October.

As you know this annual event receives generous sponsorship and publicity from local businesses, and raises funds for local and national charities. As usual we have several pro-celebrity entrants, including the jockey Frank Francis and his wife, model Junina; jazz twins Ashley and Judas Chall; and tennis prodigy Kim Ong Kwae. This year's beneficiary will be the Nottingham Hospital for Sick Children.

We would be very grateful of any coverage you could give us both before and after the event, as we are hoping to increase our attendance and charity donations this year. I have enclosed four passes that will allow you exclusive press entry to the course on both days.

It is not too late to enter teams, so if any of the Echo staff would like to take part, please call me for details.

I look forward to hearing from you.

Yours faithfully,

J W Pearce

Julian Pearce
Event Organiser

Letters to shops and businesses

Advising of a billing error

28 Smith Mansions
Howley
READING
Berks
HY5 1SC

1 December 2004

Accounts Department
Myers Wholesale Pipe Fittings Ltd
LUDLEY
Wilts
YT7 4ST

Dear Sirs

Attached please find your invoice XY76467, dated 25 November, which has been delivered to me today at the above address.

I have never ordered nor received goods from your company, nor would I have any use for them, so I can only assume this is an addressing error. Would you please check your records and remove my name and address from them.

Yours faithfully

M Cardon

Michael Cardon

Enc. copy of invoice XY76467

Complaining about faulty goods

37 Stoney Lane
CHEAM
Surrey
GH2 9WT
3 October 2004

Mr John Gladly
Gladly's Garden Products
Wenton Farm
SHIRETON
West Sussex
JK12 82G

Dear Mr Gladly,

Re: Spartan Garden Rake

On 26 September I purchased a Spartan Garden Rake from your Cheam store (see receipt attached) with the intention of clearing some flowerbeds of weeds. Unfortunately this product has not performed well as the tines bend when in contact with the soil. This means I cannot effectively rake soil in my beds.

I returned to your Cheam store to request a refund or part-exchange for another rake, but the branch manager, Mr. Jake Gurton, refused to allow this as the product showed signs of use. He suggested I address my complaint to you.

While I realise this is an 'economy' product, it should perform the task it is designed for. I am unable to use the rake effectively and believe it is not fit for its purpose. Therefore I expect a refund from either Gladly's or the manufacturer of this item.

I look forward to an early resolution of this issue.

Yours sincerely

George Grumble

Mr G H Grumble
Enc. receipt for purchase, 26/02

Letters to banks and other financial institutions

Asking for an overdraft or loan

54 Fern Road
TIPLON
West Midlands
B16 7AJ

10 June 2000

Mrs E Cox, Manager
Big Bank plc
High Street
TIPLON
West Midlands
B15 8AG

Dear Mrs Cox

Account No. 19274364

I would like to arrange for a temporary overdraft of £4,000 on the above account, with immediate effect. The money will be used for the expenses of my daughter's wedding. I am expecting a savings policy to mature in early September which will pay some £13,000 and this will be deposited into the account on receipt.

As you know I have held an account with this branch for some twenty years and have arranged such overdrafts on several occasions. These have been repaid on time and in full, so I hope you will consider this request favourably.

I would be pleased to come and discuss this matter with you, if you wish, or supply a copy of the policy in case you require it for security.

Yours sincerely

John Juston

Arranging or changing payment details

17 Hwebau Gardens
Usley
WINFORD
Bucks
TN3 5HR

12 April 2004

Policy Servicing Dept
Winford Wells Assurance Co
WINFORD
Bucks
TN4 9BR

Dear Sir or Madam,

Re: Change of direct debit for policies 7005971 (Jones D) and 7005998 (Jones F)

Please find enclosed the completed direct debit instruction to change the bank account from which monthly premiums are taken for the above life policies.

If possible, I would like you to combine this with collection of the premium for my health policy (policy number 70086592: Jones D), which is already debited from this account on 3rd of each month. That is, I would prefer to make one payment to cover all three policies, rather than have separate debits from the account on separate days.

Please would you ensure the existing direct debits for the life policies are cancelled with immediate effect.

If you have any queries regarding this matter, please telephone me on the number shown above.

Many thanks,

Doreen Jones

Doreen Jones

Letters to your landlord/tenant

Asking for repairs to be made

Flat 4
16 Holford Road
POOLFORD
Dorset
JP9 5TB
6 February 2004

Ian Graham Esq.
40 Moreley Drive
POOLFORD
Dorset
JP8 6YB

Dear Mr Graham,

Re: Urgent repairs at the above address

During the last three days a large crack has developed in the ceiling of the main bedroom in Flat 4. Areas of plaster are becoming loose and further cracks are appearing. With the tenant of that flat (Mr Jones of Flat 8) I have inspected the floor above, but we cannot find any indication of the reason for the problem.

I should be most grateful if you would arrange for a builder to inspect the property as soon as possible to establish the cause and effect any necessary work to make the ceiling safe.

Yours sincerely,

M Duttine

Michael Duttine

If there is no response to an initial letter, you can send a stronger follow-up letter. However, if an issue escalates, make sure you seek advice from your local housing centre or legal advice.

Asking a tenant to stop anti-social behaviour

19 Habger Lane
GIDLOW
Firkshire
GH6 7SF
14 April 2004

Habger Lodge
GIDLOW
Firkshire
GH6 9US

Dear Mr and Mrs Hunt

Over the last few weeks, since you moved into Habger Lodge, I have received a number of complaints from your neighbours regarding your habit of burning rubbish in the back garden at the weekend.

As you know, the property is located within a no-smoke zone, which means that bonfires are banned during hours of daylight and only wood and wood products may be burned at night. It seems you are contravening this by burning plastics, fabric and other materials in the afternoon.

No doubt the complainers have also made their point to the local council, so you may well receive a similar letter from them. As the owner of the property I am legally responsible for ensuring that you comply with these regulations, so if I receive further complaints or notification from the council that the bonfires are continuing, I shall be forced to terminate our agreement and ask you to move out.

As you are new to the district you may be unaware that there is a household waste collection point in Dellmer Avenue, which will accept all forms of household rubbish free of charge. Details are available from the site or from the local council offices.

Yours sincerely,

A Clarke

Alan Clarke

Also check out the letters to neighbours on pages 128–129.

The following more informal approach is ideal for house-shares.

26 Howton Way
HUNTSFORD
HJ87 2HW

26 November 2004

Dear Jack,

You've been living here for some six months now, and although we've had several house meetings at which I've asked you to clear up your laundry, I am annoyed to find it is still appearing in almost all the rooms in the house.

This must stop. It is not acceptable behaviour, especially as I bring business associates into my home who expect a certain standard of decorum. Last week my boss came over for drinks and was confronted by a muddy sports kit in the hallway; a pile of unwashed jeans on the kitchen counter; shirts hung to dry in the shower; and various balled socks and discarded ties in the lounge. I quickly assured her that these were not mine, but no doubt it had already made a negative impression.

If you wish to leave laundry laying about in your own room, I cannot object. But when all is said and done this is my home, and I feel that it is not my duty to collect up your clothing from around the areas we share. If the situation has not improved within one week I am afraid I shall have to ask you to look for somewhere else to live.

I'm sorry it has come to this, and hope it will not affect our friendship, but you have had several warnings and I feel my wishes are being ignored.

Regards,

John

Asking for overdue rent

A friendly, sympathetic approach often works when the reader is expecting a demand:

27 Monkton Road
High Gilford
WOLVERHAMPTON
SS76 9WG

28 January 2001

Flat 2
29 Monkton Road
High Gilford
WOLVERHAMPTON
SS76 9WG

Dear Mrs James

Re: Overdue rent

I notice from my accounts that you have not paid your rent on time for some three months, and that the amount due on 10 January is still outstanding.

I am aware that you have recently lost your job through illness, so I realise that you may be having some financial difficulties. May I suggest that you speak with the housing department or contact the Benefits Office to see whether they are able to assist you with payment? My experience with previous tenants suggests that they will, although it may not be for the full amount.

If I can help in any way with this, please do not hesitate to call me. You have been a reliable and trouble-free tenant for many years and I hope this matter can be resolved to our mutual satisfaction.

I hope your health is improving and that you are soon back on your feet again.

Yours sincerely,

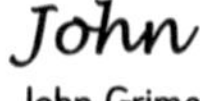

John Grimes

Explaining why rent is late/asking for more time to pay

Flat 2
29 Monkton Road
WOLVERHAMPTON
SS76 9WG

13 February 2001

27 Monkton Road
WOLVERHAMPTON
SS76 9WG

Dear Mr Grimes

I have received your two recent letters regarding the rent payment, and apologise for the delay in answering them. I have been in and out of hospital since mid-November and have rather neglected my mail. As I have been rather preoccupied with my treatment I was unaware that the rent has been unpaid for the past two months.

I am very sorry that you have received late payments, and especially that you have not yet received January's payment (for which I now enclose a cheque). Back in October, when I was told of the need for hospital treatment, I asked my bank to make these payments directly to you each month, but as my former employer has not paid my wages since my first day of sick leave the bank was unable to do so.

I have approached the Benefits Office as you suggested and they are able to help with a proportion of the rent but it will take some six weeks for the money to be paid to me. Therefore I am currently arranging a loan to tie me over this period, as I anticipate being unable to work for at least another six weeks. I hope to have these funds by next Monday, at which point I shall pay the rent for February. I hope you will allow me these extra few days' grace as I really have no other option.

I thank you for your concern and your kind suggestions. Many a landlord in your position would not have taken such a reasonable tone.

Yours sincerely,

Julia James

Mrs J James

Terminating a tenancy

Flat 2
29 Monkton Road
WOLVERHAMPTON
SS76 9WG

28 April 2001

27 Monkton Road
WOLVERHAMPTON
SS76 9WG

Dear Mr Grimes

As you know, I have been receiving hospital treatment over recent months for a muscular disorder. Sadly, my condition has not improved and my consultant now suggests that I should move into a warden-assisted flat to ensure my own safety.

I realise that this is rather short notice, a few days short of the required four weeks. I apologise for this, but I am sure you will be able to find another tenant easily. I shall be sad to leave the flat as you have been a particularly good landlord. I have spent many happy years here and have no complaints.

When I first moved in I paid a security deposit to your predecessor that was to be returned when I vacated the flat. As I shall be moving mid-month would you please deduct from this the rent due from 10–18 May and send the remainder to my new address (shown below) following your inspection of the flat on my departure. I am sure you will find the property to be in an acceptable and clean condition.

Many thanks for all your help over the past 17 years.

Yours sincerely,

Julia James

Mrs J James

New address from 18 May: Mrs J James, 101 St Swithin's Court,
WOLVERHAMPTON SS76 7ST

When you, as landlord, decide to move your tenant on:

27 Monkton Road
WOLVERHAMPTON
SS76 9WG

18 February 2001

Flat 2
29 Monkton Road
WOLVERHAMPTON
SS76 9WG

Dear Mrs James

Re: Overdue rent

I have now written to you on two occasions regarding the late and non-payment of your rent. To date, I have received no response from you. I have also telephoned and visited the property several times without success.

As indicated in my previous letter, I therefore have no alternative but to give you notice of the termination of your tenancy. You are required to vacate the premises in four weeks' time, that is, on or before the 18 March. Your security deposit will not be returned as it will be retained in lieu of the rent owed. In acknowledgement of your health and financial problems and your previous good record of payment I shall not pursue you for the remaining debt if you vacate the premises on time.

I am sad that we have to end our long association in this way, but I have given you every opportunity to discuss the matter and have suggested ways in which you may be able to obtain financial help, but these have been ignored. I therefore have no alternative but to seek another tenant who is able to pay the rent on time.

I hope your health improves and that you find a suitable alternative property.

Regards,

J Grimes

John Grimes

Letters to your employer

Asking for time off

There may be many reasons why you need time off. For example:

> 4 Farren Road
> LONDON
> SE6 8UH
>
> 2 July 2004
>
> Mr Nichols
> Sales Manager
> Grant & Sons Ltd
> 19 Church Lane
> LONDON
> SE4 91G
>
> Dear Mr Nichols
>
> I am writing to request your permission for me to be absent from work next week (8–12 July).
>
> I am needed to look after my invalid mother during this week as the carer who normally looks after her has been taken ill, and at such short notice I have been unable to arrange alternative care. This is an exceptional situation, which I do not anticipate recurring.
>
> I appreciate that I would not be paid for this week.
>
> Yours sincerely
>
> *J. Ryman*
> Ms Jane Ryman

Many employers would be happy to authorise such leave on the basis of a verbal request, but it is always best to provide a written request and request a written answer, in case of any future dispute over the matter.

The next example is a letter excusing absence from work because of the writer's own illness.

15 Greenland Road
SAMPTER
Essex
ES3 8UJ

3 December 2000

Mr H Jones
Head of Sales Department
Hampton & Co Ltd
12-14 High Street
SAMPTER
Essex
ES7 4ED

Dear Mr Jones

I left a telephone message with your secretary this morning to explain that I was feeling unwell, and was going to visit my doctor.

He has diagnosed an infection of the respiratory tract, and suggested I would probably need to be off work for at least ten days. I enclose a medical certificate.

I have spoken to John Griffiths about my work commitments for the next few days and I believe that things are under control, but if anybody has any problems about my work they should not hesitate to telephone me.

Yours sincerely

Gerry

Gerald Long

Enc.

Enquiring about promotion prospects

2 Hornsby Terrace
UPTON
Surrey
SU8 3WE

5 June 2000

Ms K Tope
Accounts Manager
Smith & Co. Ltd
54-58 Pine Road
UPTON
Surrey
SU4 5RF

Dear Ms Tope

I am writing to ask whether you would consider promoting me to the position of Senior Clerk.

As you know, I have been working for the company for four years, one year as Junior Clerk and three years as Clerk. I have undertaken the work of Senior Clerk during periods of holiday and sickness and, apart from finding the work very interesting, I believe that I have performed it satisfactorily. I have a number of ideas for improving the efficiency of the Department, which the scope of my present job does not allow me to implement.

I would be grateful for the opportunity to discuss this matter.

Yours sincerely

Rob Smith

Robert Smith

Resigning from your job

The following letter could be adapted for most situations.

4 Oak Road
ALSOP
Yorks
NO7 3SD

29 September 2000

Mr D Hobbis
Sales Manager
Creen & Sons Ltd
25 Weldon Road
LEEDS
LE1 8NM

Dear Mr Hobbis

I have been offered, and have decided to accept, the position of Sales Manager with Broom & Sons Ltd. I am writing, therefore, to give you the appropriate four weeks' notice to terminate my employment with the Company on 27 October.

I have been very happy during my years here and it was with some regret that I reached this decision. However, my new position offers considerably more scope and responsibility than my present one.

I would like to take this opportunity to thank you for all the support and guidance you have given me over the past five years.

Yours sincerely

Frank Rane

Frank Rane

Complaining to your employer

If you are in a union you might wish to write to them, or send them a copy of the letter.

Accounts Department
Third Floor

6 February 2000

Ms D Kenny
Personnel Manager
Fifth Floor

Dear Ms Kenny

I wish to draw your attention to the washing facilities on the third floor. Only one sink is provided for over one hundred male employees on this floor. There is usually neither soap nor clean towels available.

This must be detrimental to the health of employees and is certainly a waste of the Company's money since much work time is lost in queuing for the use of these inadequate facilities. I believe this, in fact, to be in contravention of the laws relating to Heath and Safety At Work.

I would be grateful if you would look into this matter urgently, with a view to providing adequate, well-maintained facilities.

Yours sincerely

John Harris

John Harris
On behalf of the Accounts Department

Letters about jobs

Asking for work

18 Field Gardens
BRIMTON
Beds
MB6 5PR

7 October 2000

Mr M Davies, Staff Manager
J Bloggs & Sons Ltd
BRIMTON
Beds
MB4 PQ8

Dear Mr Davies

I am writing to enquire whether you have any vacancies for a junior office assistant.

I left King Edward's School this summer, having passed GCSE Examinations in English Language, Geography and French. In my last year at school I also took a course in typing and general office procedures, and I am keen to make proper use of these skills.

Since leaving school I have worked as a filing clerk for Smiths Caterers, providing maternity cover for their usual clerk who is now returning to work. Mr Smith will be quite willing to give me a reference.

I realise that you may not have any vacancies at present, but I would be willing to work in any department and to undertake any training required whenever a possible position arises.

I look forward to hearing from you.

Yours sincerely

C. G. Brown

Caroline Brown (Miss)

Asking for details of an advertised job

15 Oakfield Drive
CARTOWN
Hampshire
L12 7RT

15 January 2004

The Secretary
The Empire Trading Company
Ships Wharf
BRIARLEY
Hampshire
SO5 7RX

Dear Sir or Madam

Further to your advertisement for an Export Manager in The Briarley Echo (9 January 2004), I would be grateful if you would send me an application form and further details of the position, including any experience or qualifications required.

I look forward to hearing from you.

Yours faithfully

F J Binns

Frank Binns

Asking a new acquaintance for help

26 Rowley Way
HOWTON
West Yorkshire
WF8 6XJ

12 July 2003

Dear Mrs Jennings

Some weeks ago we were introduced at Kevin and Sharon's wedding, and had a long chat about working in childcare. I'd like to thank you for giving me some great ideas and explaining the options for getting qualified. It was very helpful and has made me more determined to look for a suitable position.

You mentioned that you may know someone who would be interested in taking on new staff. Would you consider giving me their details, or passing on my enclosed CV to them? I would very much appreciate this – I'm sure you realise how difficult it can be to get your foot in the door without recommendation.

It was very enjoyable to meet and talk with you, and I hope our paths cross again in the near future. Meanwhile, thank you again for all your help and advice.

Regards

Julie

Julie Johnson

Applying for the job (no CV)

55 Davidson Avenue
MANTOWN
MN7 7TS

21 January 2003

Mr H Ody
The Apsley Motor Company
Jowett Road
MANTOWN
MN3 IAK

Dear Mr Ody

I wish to apply for the position of skilled motor mechanic as advertised in today's Mantown Gazette.

Having completed a full apprenticeship, I have been employed by Jones Motors in the London Road for the past two years. As Mr Jones is retiring shortly and closing the business, I am anxious to move to a firm where there are both prospects for promotion and interesting, varied work.

Mr Jones has agreed to give me a reference, and I would be free to attend an interview at any time.

I look forward to hearing from you.

Yours sincerely,

John Collins

John Collins

Accepting a job offer

Ivy Cottage
Compton Road
NACWICH
Berks
SL5 6XC

15 October 2004

R Burns Esq
Personnel Manager
Timetec Limited
Rose Estate
OXRIDGE
Bucks
SL9 5RF

Dear Mr Burns

Thank you very much for your letter of 12 October offering me the position of Sales Representative with your company.

I am delighted to accept the position, and look forward to starting work with you on 13 November.

Yours sincerely

June Alcock

Miss J Alcock

Rejecting a job offer

Ivy Cottage
Compton Road
NACWICH
Berks
SL5 6XC

15 October 2004

R Burns Esq
Personnel Manager
Timetec Limited
Rose Estate
OXRIDGE
Bucks
SL9 5RF

Dear Mr Burns

Thank you very much for your letter of 12 October offering me the position of Sales Representative with your company.

However, I regretfully have to decline as I have been offered, and have accepted, a similar position with a company that is much closer to my home. As I have a young family, I wish to spend as much time with them as work commitments will allow and reducing my travelling time will help considerably.

Thank you once again for offering me the position, and for your confidence in my abilities.

Yours sincerely

June Alcock

Miss J Alcock

Letters of reference

Requesting a character reference

4 Rose Wood
OXRIDGE
Bucks
SL5 6XC

15 February 2004

Mr Fred Jones
Projects Director
Timetec Limited
OXRIDGE
Bucks SL9 5RF

Dear Mr Jones

I am applying for a mortgage from Big Bank, and I wondered whether, as a senior representative of my employer, you would provide me with a written reference to support my application.

Such a reference would contain details of my current position and my recent promotions, current salary and future salary expectations, and an assessment of my character. The reference should be sent directly to the bank at the address shown below, quoting the reference shown.

If you are unable to do this, would you please let me know so I can ask someone else.

Yours sincerely

Janice Godsell

Mrs J Godsell

Bank address: Big Bank, 30 High Street, OXRIDGE SL9 3HY

Please quote ref: M293726/JWG.

Requesting an employment reference

25 Oakley Avenue
MILTON
Northants
SY12 4LT

14 May 2004

Mr B Brown
M & B Creative Marketing
Dower House
PORTON
Northants
RT6 3ER

Dear Mr Brown

I am applying for the post of Salesman with Kingley Marketing of Milton, and I wondered whether you would be willing for me to give your name as a referee.

I have been very happy in my present post, as I was during my four years with M & B, but I have decided to apply for the post with Kingley Marketing as it seems to offer greater responsibility and a chance to use my own initiative more frequently.

Please pass on my regards to any of my colleagues still with M & B and, of course, to Mrs Brown.

Yours sincerely

Brian

Brian Butts

Giving a reference

2 Clarion Close
HENLEY
Surrey
WN8 9GY

10 March 2004

Personnel Manager
Get Mobile Ltd
Henley Industrial Park
HENLEY
WN8 6TV

Dear Sir or Madam

Re: Jane Golding

I have worked with Jane Golding since she started with Fones R Us in 1997. In that time she has progressed from office junior to call centre supervisor.

Jane's responsibilities include training staff, handling contentious calls and resolving customer complaints. She has always expressed herself with good humour and tact and has received several personal thank-yous from clients. She has been the recipient of our 'employee of the month' award on several occasions and her attitude and achievements have attracted favourable comment from our CEO.

I am happy to confirm that Jane is of good character – she is both reliable and honest – and is very easy to get along with. She is a motivator and is happy to organise staff social events as well as giving extra time and assistance to those who need it in their work. I am sure she will be an asset to your organisation, although we shall be sad to lose her.

Yours faithfully

P J Young
Paul Young
Manager – Call Centres

Chapter 12

Looking for extra help?

Still short of ideas or stumped by a particular situation? Here are some books and web sites you might like to look at for further inspiration, and organisations that may be able to help with difficult situations.

> If you don't have access to the internet, ask at your local library. Most offer limited use of the internet free or for a small fee, and may offer tuition or advice on getting started. Some also have word processing software you can use to write your letters, and allow you to print them for a minimal charge.

Tips for better letters

The following are great all-round guides to writing letters:

- *Chambers Guide to Letter Writing* ed. Kay Cullen (Chambers: ISBN 0550141308)
- *Great Letters for Every Occasion* by Rosalie Maggio (Prentice Hall: ISBN 0735200815)
- *E-mail and Business Letter Writing* by Lynn Brittney (Foulsham: ISBN 0572025793)
- *Janner's Complete Letter Writer* by Greville Janner (Random House: ISBN 0091740681)

If you need to write to Americans, try *Everyday Letters for Busy People* by Debra Hart May (Career Press: ISBN 1564143392). It contains tips on structuring letters, some great samples and information on writing to American public figures.

The following web pages contain valuable tips on particular aspects of letter-writing, or links to useful sites:

Business letters:
www.ipl.org/div/pf/busletters.html, www.business-letters.com, www.4hb.com/letters/index.html

Letters to public figures:
(UK) www.debretts.co.uk/people/address.asp
(American) www.cftech.com/BrainBank/OTHERREFERENCE/FORMSOFADDRESS/ SpkWritFrmsAddr.html

Job applications and CVs: www.bradleycvs.co.uk/jobhunt/letters.htm, www.careerlab.com/letters

Penpals: www.worldkids.net/girl (for girls aged 8-14)

Letters/petitions on human rights and associated issues:
www.amnesty.org, www.stoptorture.org

Love letters:
www.sparks.com/valentines_day/ehow_write_love.html
www.electpress.com/loveandromance
www.geocities.com/Paris/LeftBank/4528/letters.htm

Wedding invitations and thank-yous:
www.usabride.com/wedplan/t_invitations.html

How to e-mail

See *The Beginner's Email Book* by Helen Smith (Foulsham: ISBN 0572026749)
For help with smileys (emoticons) try: www.webopedia.com/smiley, www.freecomms.co.uk/Various/smile.asp
For information on netiquette try: www.whatis.com/netiquette

Help with texting

For deciphering all those weird abbreviations, try:
Text Me (Penguin: ISBN 0670910791)
The Total TxtMsg Dictionary by Andrew John and Stephen Blake (Michael O'Mara Books: ISBN 1854798936)
The Txtrs A-Z also by Andrew John and Stephen Blake (Michael O'Mara Books: ISBN 1854798634)

For all you need to know about texting, including how to, fun stuff, abbreviations and smileys, try: www.text.it, www.smsportal.com
The sites listed under e-mail for smileys are also useful.

Junk mail and spam

Mailing Preference Service
DMA House, 70 Margaret Street, London W1W 8SS
Tel: 020 7291 3310/Fax: 020 7323 4226
Web: www.mpsonline.org.uk (mail)
or www.e-mps.org.uk (e-mail and texts).
The DMA also run similar lists to stop nuisance telephone marketing or fax marketing calls.

Advertising Standards Authority
2 Torrington Place, London WC1E 7HW
Tel: 020 7580 5555/Fax: 020 7631 3051
Email: inquiries@asa.org.uk or web: www.asa.org.uk

A complaints form is available on the web site, and it is recommended that you send a copy of the offending advertisement with this or your written letter.

The Information Commissioner
Wycliffe House, Water Lane, Wilmslow, Cheshire SK9 5AF
Tel: 01625 545745/Fax: 01625 524510
Email: data@wycliffe.demon.co.uk or data@dataprotection.gov.uk
Web: www.dataprotection.gov.uk

Online resources for spam: www.spamcop.com, www.stopspam.org.uk

Consumer issues

Which? (The Consumers' Association)
Castlemead, Gascoyne Way, Hertford X SG14 1YB
Tel: 08453 010010
Fax: 020 7770 7485
Web: www.which.net

BBC Watchdog
201 Wood Lane, London W12 7TS
Tel: 0870 010 7070
Web: www.bbc.co.uk/lifestyle/watchdog

Media issues

Independent Television Commission
33 Foley Street, London W1W 7TL
Tel: 020 7255 3000
Web: www.itc.org.uk

Press Complaints Commission
1 Salisbury Square, London EC4Y 8AG
Tel: 020 7353 1248
Web: www.pcc.org.uk

Broadcasting Standards Commission
7 The Sanctuary, London SW1P 3JS
Tel: 020 7233 0544
Web: www.bsc.org.uk

Index